T0219842

A Whirlwind History of the Universe and Mankind

Thomas Sanford

A Whirlwind History of the Universe and Mankind

From the Big Bang to the Higgs Boson

 Springer

Thomas Sanford
Department of Diagnostics
and Target Physics
Sandia National Laboratories
Albuquerque, NM, USA

ISBN 978-981-97-2673-8   ISBN 978-981-97-2674-5   (eBook)
https://doi.org/10.1007/978-981-97-2674-5

This Springer imprint is published by the registered company Springer Nature Singapore Pte Ltd.
The registered company address is: 152 Beach Road, #21-01/04 Gateway East, Singapore 189721,
Singapore

Paper in this product is recyclable.

*To my Father
F. Bruce Sanford
my early teacher who
showed me the beauty
of science*

and

*To my later guide
Leon M. Lederman
who taught by example*

Preface

This book is written for the person who is curious about how we humans came to be here and who is interested in understanding a little of the science and social evolution that enabled us to affirm that a Big Bang actually happened and for the need of a Higgs Boson.

The text emphasizes, where possible, the why behind various evolutionary plateaus, from the universe's beginning in the Big Bang, to the visible heavens, our solar system and to Earth, with its evolution of life. From the emergence of Homo sapiens and the agricultural revolution, the discussed history narrows from that of the early Middle East, to the development of the Mediterranean civilizations, including Greece and Rome, to the European Renaissance, the English industrial revolution, and to the early European science discoveries, particularly those in physics. This path, through to the American Manhattan Project, leads to the exploration of what is inside the nucleus in the new field of high-energy particle physics. It is this road, which gave rise to theorizing the existence of the mass-giving Higgs boson and then discovering it, that is articulated; all condensed in less than 200 pages and fulfilling the book's title.

In the book's beginning, known science is used to explain the evolution from the Big Bang. Later, as Homo sapiens became humans and society developed, this early history and eventually the science that evolved the physics necessary to understand the beginnings is explained historically. Only the present scientific consensus from a microsecond after the Big Bang is discussed. Prior to the Big Bang, no understanding based on science is known to the author nor is commented on. A summary of the major events, during each of the three time periods discussed (beginnings, humans, and physics), is included as a time-line in the Key Events Discussed. References consulted are listed in the Bibliography at the end of the book.

In the text, a few equations are used to illustrate the concepts being discussed. These are there to illustrate the innate beauty and simplicity of the ideas behind the words. The equations are easily skipped without loss of understanding of that which is written. Although the book is meant to be read as a whole, the material covered is vast. However, each of the chapters is self-contained and can be read as a unit without loss of comprehension.

To assist in guiding understanding of the unfamiliar places mentioned, insightful maps are easily available on the Web, which the author suggests using. Similarly, technical terms are occasionally used in the various areas of science discussed and these are also quickly explained by referring to the Web.

Credibility to that which is written is given by the author's eight years of research in high-energy particle physics at both the Nevis (Columbia University) and Brookhaven National Laboratories in the USA, seven years at the Rutherford High-Energy Laboratory in England and CERN (French: *Conseil European pour la Research Nucleaire*), Laboratory in Switzerland, his nearly two decades of research into pulsed power technology and nuclear fusion as a Distinguished Member of Technical Staff at the Sandia National Laboratories in NM and his travels throughout the regions discussed (except that of Iran and Iraq). In 1973, the author received his Ph.D. under Nobel Laureate Leon M. Lederman. In 2000, he became a fellow in the American Physical Society, and in 2005, he received the Hannes Alfven Prize by the European Physical Society. He is the lead author on 60 of 102 journal publications and author or contributor to 281 technical articles. A short summary of his science background can be found in the author's *History of HERMES III Diode to Z-Pinch Breakthrough and Beyond (learning about Pulsed Power and Z-Pinch ICF)*, published by Sandia National Laboratories as SAND2013-2481, April 2013, and available on the Web.

Albuquerque, USA   Thomas Sanford

Acknowledgements

To the many friends and colleagues who have encouraged this work, the author is grateful. In particular, the following are appreciably acknowledged: Dotty Noe who made useful editorial comments, during the early phases of the writing, Thomas D. Sullivan and Drs. Susan L. Harper, Hans B. Jenson, David J. Johnson, Nino R. Pereira, Kenneth W. Struve, Mary Ann Sweeney, and A. J. Toepfer for their detailed suggestions whose edits enhanced the clarity of the presentation. A. J. Toepfer and Hans B. Jenson are especially thanked for the extra accuracy brought to the physics parts and the high-energy physics chapter, respectively. The sustained encouragement by Kenneth R. Prestwich, Dr. Mary Ann Sweeney and the author's partner Dr. Susan L. Harper, and Drs. James R. Asay, Yasuyuki Horie, and Akiyuki Tokuno were essential for promoting the book's publication and for the completion of this eight-year project.

Contents

Part I Beginnings

1 Introduction 3

2 Matter Universe 5
Early Big Bang 5
Galaxy Evolution 7
Star Evolution 8
Solar System Evolution 10
Earth-Moon Evolution 11
Earth Evolution 12

3 Life 15
Primitive Life and Snowball Earth 15
    Prokaryotes 15
    Great Oxidation Event and First Snowball Earth 16
    Eukaryotes 17
    Three Additional Snowball-Hothouse Cycles 18
Complex Life and Mass Extinctions 19
    Ediacaran Period (635–541 MYA) 19
    Cambrian Period (541–488 MYA) 20
    Ordovician Period (488–444 MYA) 20
    Silurian Period (444–416 MYA) 20
    Devonian Period (416–359 MYA) 21
    Carboniferous Period (359–299 MYA) 22
    Permian Period (299–252 MYA) 22
    Triassic Period (252–200 MYA) 23
    Jurassic Period (200–145 MYA) 24
    Cretaceous Period (145–66 MYA) 24
    Cretaceous-Paleocene Mass Extinction (66 MYA) 25
    Paleocene Epoch (66–56 MYA) 26

Eocene Epoch (56–34 MYA) 27
Oligocene Epoch (34–23 MYA) 27
Miocene Epoch (23–5.3 MYA) 28
Pliocene Epoch (5.3–2.6 MYA) 29
Pleistocene Epoch (2.6 MYA-11,700 YA) 29
Holocene Epoch (11,700–0 YA) 31

Part II Humans

4   Human Evolution 35
Apes to Homo Sapiens (8–0.2 MYA) 35
East Africa, Chimpanzees and Hominins 35
Hominin Species 36
Need to Adapt 37
Out of Africa Hypothesis 38
Hunter-Gatherer (200,000–10,000 YA) 39

5   Agricultural Revolution 41
Transition to Agriculture 41
Early Villages 42
Jericho 43
Catalhoyuk 43
Gobekli Tepe 43
Continued Development 44
Early City-States in Sumer 45
Early Empires 47
Other Early City-States 48
Nile River Valley 49
Indus River Valley 51
Yellow and Yangtze River Valleys 53
Other Regions 53
Advantages of Old World Over New World 54

6   Mediterranean Development 57
Mediterranean Physical History 57
Migrations Around the North and Eastern Mediterranean Basin 59
Early City-States in the Eastern Mediterranean Basin 60
Minoans 61
Mycenaeans 61
Greeks 62
Phoenicians 63
Greece, Persia and Hellenization 64
Evolution of Greek Thinking 66
Greek Contributions 66
Hellenistic Contributions 69

Rome: The Evolutionary State in the Western Mediterranean Basin 70
   Rome's Beginnings 71
   Rome's Expansion Beyond Italy 72
   End of the Republic 74
   Early Roman Empire 75
   Roman Empire's Golden Years 76
   Roman Empire's Declining Years 78

7   **Europe's Beginnings** 87
   Early Middle Ages: 500–1000 (Kingdoms) 87
      Barbarian Kingdoms 87
      The Franks 89
      Predecessors of France and Germany 91
      Britain Becomes England 91
      Spain Becomes Al-Andalus 93
   High Middle Ages: 1000–1300 (Recovery from the Dark Ages) 94
      Improvements 94
      Towns 94
      Clocks 96
      Woolen Cloth 96
      Gothic Cathedrals 97
      Scholasticism 97
      Universities 98
      Crusades 99
      Investiture and Kingdom of Germany 100
      Magna Carta and Parliament 101
      Marco Polo 102
   Late Middle Ages: 1300–1500 (Challenges) 103
      Avignon France 103
      The 100 Years War 103
      The Black Death 104
      Gunpowder 105
      Printing Press 106
      Italian Renaissance 107
      Byzantine Empire Ends 112
      Draw of Spices and Gold 112

8   **Transition: 1500–1700 (The Known World Expands)** 115
   Voyages of Discovery 115
      Christofer Columbus 115
      Amerigo Vespucci and Martin Waldseemuller 116
      Nunez de Balboa 117
      Ferdinand Magellan 117
   New Exchange Networks 118
      The Columbian Exchange 118
      The First Global Exchange 119

The Atlantic Exchange 120
The Reformation and Religious Wars 120
The Scientific Revolution 123
  Scientific Method 123
  Francis Bacon and Rene Descartes 128
  Royal Society of London 128
The Enlightenment 129
  John Locke 129
  Baron de Montesquieu 130
  Voltaire 130
  Denis Diderot 130

9   Modernity: 1700–1900 (Industrialization and the Steam
     Engine) 133
     English Industrial Revolution 133
       Preconditions 133
       Textiles 134
       The Steam Engine 135
       Birth of a New Society 137
     Industrialization Spreads 138
       Western Europe 138
       United States 138
     Empires Late to Industrialize 139
       Russia 139
       China 140
       India 140
       Ottoman 141
     Two Revolutions and Two Canals 142
       American and French Revolutions 142
       Suez and Panama Canals 143

Part III   Physics

10   Physics: 1700–1900 149
     Thermodynamics 149
       Energy 149
       Heat 150
     Atoms, Elements and Molecules 151
     Temperature and the Ideal Gas Law 152
     Light 153
     Electrodynamics 153
       Electricity 154
       Magnetism 155
       Electromagnetic Waves 156
     Additional Important Discoveries of the Nineteenth Century 158

11 Physics: Around 1900 161
   X-Rays 161
   Radioactivity 162
   The Electron 163
   Quantum Structure 165
     Blackbody Radiation 165
     Photoelectric Effect 166
   Space and Time 167
     Special Theory of Relativity 167
     General Theory of Relativity 170

12 Physics: 1900–1950 173
   Atomic Structure 173
     Rutherford Model 173
     Bohr Model 175
     De Broglie Model 176
     Electron Spin 177
     Schrödinger and Heisenberg 178
     The Neutron 180
   Nuclear Discoveries 181
     Nuclear Fission 181
     Nuclear Chain Reaction 182
     Nuclear (Atom) Bomb 183
     Thermonuclear (H) Bomb 184
     Nuclear Beta Decay and the Neutrino 186
     Nuclear Sun and Solar Neutrinos 187
   Quantum Electrodynamics (QED) 189

13 Particle Physics: 1950–2023 191
   Cosmic Rays 191
   Symmetry Violations and Family-Concept 193
     Pions and Muons 193
     Parity Violation 193
     Two Different Neutrinos 195
     Antiprotons and Antineutrons 197
     Charge Parity Violation 198
   Hadrons and Quarks 199
     The Bubble Chamber 199
     New Resonance States 200
     The Quark Model 200
   More Quarks and Leptons 202
     Charm Quark 202
     Tau Lepton 204
     Bottom Quark 205
     Top Quark 206

Gluons 208
Intermediate Vector Bosons 209
    Prediction 209
    Indication at Gargamelle 210
    Discovered at SPS Collider 210
    Detailed Measurements at LEP 212
The Higgs Boson 213
    Electroweak Problem 214
    Solution 214
    The Standard Model 216

14 Summary 219

Epilogue 221

Key Events Discussed 225

Bibliography 233

Abbreviations

AGS	Alternating Gradient Synchrotron
BNL	Brookhaven National Laboratory
CERN	French: *Conseil European pour la Research Nucleaire* (European Council for Nuclear Research)
CGS	Centimeters, Grams, Seconds
CMB	Cosmic Microwave Background
CNTBT	Comprehensive Nuclear-Test-Ban Treaty
COBE	Cosmic Background Explorer satellite
CP	Charge Parity
CPT	Charge Parity Time
DESY	Deutsches Elektronen-Synchrotron
DNA	Deoxyribonucleic Acid
DORIS	Eelectron-Positron Collider at DESY
GPS	Global Positioning System
GTR	General Theory of Relativity
HEP	High-Energy Particle Physics
IBM	International Business Machines
LEP	Large Electron-Positron Collider
LHC	Large Hadron Collider
LIGO	Laser Interferometer Gravitational-Wave Observatory
MCS	Multiple Coulomb Scattering
MIT	Massachusetts Institute of Technology
MKS	Meter, Kilogram, Second
NAL	National Accelerator Laboratory
NASA	National Aeronautics and Space Administration
PETRA	Positron-Elektron-Tandem-Ring-Anlage Collider at DESY
PS	Proton Synchrotron
PTBT	The Partial Test Ban Treaty
QED	Quantum Electrodynamics
RCO	Roman Climate Optimum
RNA	Ribonucleic Acid

SLAC	Stanford Linear Accelerator Center
SM	Standard Model of Particle Physics
SPEAR	Positron-Electron-Asymmetric-Ring
SPQR	Senatus Populusque Romanus
SPS	Super Proton Synchrotron
STR	Special Theory of Relativity
USA	United States of America

Part I
Beginnings

Chapter 1
Introduction

Amazingly, our universe had a birthday 13.7 billion years ago (BYA). Amazingly, we have a credible scientific explanation of how we have come to know this. And, even more amazingly, we live in a time that enables us to tell this story.

This book provides a brief timeline of many of the key events between the universe's birth, now referred to as the Big Bang, and the recent discovery of the Higgs boson. Interaction with the boson's quantum field gives mass to us all, providing credibility to the physics of the Big Bang itself.

A century ago the known universe consisted solely of the Milky Way galaxy, the Sun and our planets. Only in 1929, did Edwin Hubble (1889–1953), using the new Mount Wilson telescope, discover that our Milky Way was not alone in the universe but was one of billions of other galaxies. Moreover, these other galaxies were all moving away from the Earth. The farther away these galaxies were, the faster they moved, as if they originated in a Big Bang. But it was not until 1960, when the Cosmic Microwave Background (CMB) predicted by Big Bang theories was accidentally observed, that scientists believed an actual Big Bang had occurred. It was only then that the Big Bang was taken seriously.

The identification of the Big Bang less than 100 years ago was just one of a series of discoveries that have altered our understanding of the world in which we live. Examples of other dramatic and relatively recent scientific discoveries are the following:

- DNA—the basis of life—was discovered in 1953.
- Plate tectonics—the movement of continents—was verified in 1960.
- Tiktaalik—the link between fish and quadrupeds like ourselves—was found in 2004.
- The Higgs boson—which gives all of us, including the world, mass—was discovered in 2012.
- Gravity waves—from the collision of two black holes—was observed in 2015.
- An actual image—of a black hole—was constructed from its surroundings in 2019.

© The Author(s), under exclusive license to Springer Nature Singapore Pte Ltd. 2024
T. Sanford, *A Whirlwind History of the Universe and Mankind*,
https://doi.org/10.1007/978-981-97-2674-5_1

The formation of a black hole is like an inverse of the Big Bang in that all matter within its vicinity is gravitationally sucked into its interior and turned back into energy. No light is ever able to escape.

Such fundamental observations, together with an increasingly detailed historical record, makes it possible to provide a relatively complete story of how we came to be who we are. Human social development, with its collective learning, and the concurrent evolution of the sciences, with its associated technologies, have been essential in enabling this understanding.

Much of this knowledge is summarized in books known as *Big History.* Many contain beautiful illustrations and explanations that describe the sequence of events that have occurred throughout the world and that have led to today's awareness. An excellent example of this approach is David Christian's *Maps of Time: An Introduction to Big History,* published 2005 and his expanded book *Big History Between Nothing and Everything,* published 2014, together with two coauthors.

Motivated by a similar perspective, the author of this book focuses here on the science and the relevant social and cultural history. The book limits itself to developments in the West, where science and technology first took root. By keeping the discussion centered, this approach generates a historical timeline that can be understood and retained. To make the discussion plausible, as well as interesting, a sufficient amount of detail is included.

To achieve these goals, the book is separated into three parts and fourteen chapters. Part I discusses the generation of the physical universe from a millisecond after the Big Bang and is referred to as BEGINNINGS. It includes the formation of galaxies, stars, the Earth and the evolution of early life on Earth. Part II, HUMANS, begins when the first hominins evolved into Homo sapiens and developed social structures leading to the development of science. It includes the agricultural revolution, early city-states in the Middle East, empires in the Mediterranean Basin, in particular, Greece and Rome, and the formation of the European states to the French revolution. Part III, PHYSICS, includes the development of Western physics in the eighteenth through twentieth centuries. These chapters discuss the historical understanding of the basic constituents of matter formed during the Big Bang, thereby providing credence to the model for the Big Bang itself. The book closes with the discovery of the Higgs boson, whose quantum field gives mass to ourselves and the universe.

Chapter 2
Matter Universe

Early Big Bang

Extrapolating the measured mass density of the universe to near a time when the universe's mass converged 13.7 BYA, suggests temperatures and densities reached unimaginably high values. Extrapolation of the radiation (photon) density of the universe to this time indicates that the radiation-energy density dominated the mass-energy density in an explosive fire ball.

Within a microsecond of this beginning, as the universe continued to expand and cool, the mass density and temperatures decreased to conditions that are now studied in particle accelerators. These studies suggest that a millisecond after the Big Bang, mass in the form of quarks and leptons, and the force carrying particles, bosons, were easily produced from the radiation itself. The lowest mass leptons are the electron and neutrinos. The Higgs boson gives particles their mass through interaction with its associated quantum field. Typical processes included pair production, where collision of two photons generates a particle and an antiparticle pair. The result was the formation of a plasma sea of quarks, leptons and bosons within a radiation field of photons. (The unfamiliar terms, such as plasma, quarks, leptons, neutrinos, etc., are all explained in detail in later sections of the book as their discovery and development evolved historically.)

As this plasma continued to cool, the quarks bound together to form numerous elementary particles, which themselves decayed into other elementary particles and radiation. After 1–100 s had elapsed, the only discrete particles remaining were the protons (formed from two up and one down quark), neutrons (formed from one up quark and two down quarks), electrons, neutrinos and the ever present Higgs field.

At this time, the neutrons and protons were in thermal equilibrium, converting back and forth into one another through a weak nuclear processes; as in electron-proton collisions, which form a neutron and neutrino, and vice versa. Because a neutron weighs more than a proton and electron combined, more energy is required to make the conversion from proton to neutron than vice versa. Thus, as the universe

T. Sanford, *A Whirlwind History of the Universe and Mankind*,
https://doi.org/10.1007/978-981-97-2674-5_2

continued expanding and cooling, the number of energetic electrons needed to make the conversion decreased.

Approximately one second after the Big Bang, the rates of conversion became too slow to keep up with the expansion, so the proton-to-neutron conversion stopped. An isolated neutron is unstable and decays into a proton, electron and neutrino with a lifetime of fifteen minutes in a process referred to as nuclear beta decay.

Initially, there were enough neutrons to make the isotopes of hydrogen nuclei: deuterium, which contains one neutron and one proton, and tritium, which contains two neutrons and one proton. Eventually, nuclei of helium, containing two bound protons, and a tiny amount of lithium, which contains three bound protons, were produced through the process of nuclear fusion. By about three minutes, however, the excess neutrons were used up before elements with higher numbers of protons in their nucleus could be made.

Protons are the nuclei of hydrogen. Today, this simplest form of matter comprises 70% of all stable observable matter in the Universe, while helium accounts for 27%. The higher atomic-number elements, which now constitute the remaining 3%, appeared later, during the formation of stars. Neutrons are required to stabilize such nuclei. The strong force, which binds the nuclei of elements together, becomes overwhelmed by the repulsive electromagnetic force created by the addition of positive protons to the nucleus, during the short-time available for nuclear fusion to take place.

During this early time, radiation dominated over matter. As soon as the nuclei tried to form atoms with the available electrons, the fierce radiation field destroyed them. The cosmos remained a randomized and structureless plasma in which the radiation and matter were in equilibrium with one another.

As the universe continued expanding and cooling, the radiation-energy density decreased faster than the mass-energy density. Eventually, more than a thousand years after the Big Bang, nuclei began to attract electrons and formed atoms. The electromagnetic force between the positive protons and negative electrons pulled them together. The decreasing energy of the radiation no longer had the strength to break the atoms apart. At this time, matter emerged as the principal component of the Universe. Radiation decoupled from matter and the universe became nearly transparent, with most photons traveling unhindered through space. This transition occurred when the universe reached an age of approximately 0.5 million years (MY) after the Big Bang.

This radiation cooled and its wavelength was red shifted (stretched) as the universe continued expanding. Today, it is detectable as the cosmic microwave background (CMB) radiation, which fills the universe around us as an ancient hold-over from the Big Bang. It's spectrum agrees precisely with that expected from a blackbody radiation source, which is defined as radiation that has been thoroughly randomized and is in thermal equilibrium with its environment. This assumption of homogeneity and near isotropy is a key component in the Big Bang modeling and gives credence to the model itself.

Summarizing, the recession of the galaxies suggested that at some past time the universe began as a compact, hot, expanding entity. The observation of the CMB radiation, having a measured blackbody spectrum, was the most compelling and supporting evidence for the Big Bang. The CMB radiation relied on a different set of physical processes for its explanation than did the cosmological explosion theory for the Big Bang.

Additionally, the measured abundance of hydrogen and helium, with little else in the present universe, independently supports the Big Bang model. The universe cooled so fast, during its early expansion, that little time was available for all but the simplest nuclei to form. Moreover, no object in the universe has been found to be older than 13.7 billion years. Telescopes that observe the universe, as it was 10 billion years ago, show the early universe looked different than the universe appears today. That early universe was more crowded and contained objects like quasars that are now rare. A quasar (quasi stellar radio source) is a distant system powered by a super massive black hole billions of times more massive than our sun. Accordingly, the universe is not in a steady state but rather has changed significantly over time. This observational evidence strongly supports the Big Bang model. The Big Bang generated the space, time, energy and matter world in which we live today. Galaxies, stars and our Earth formed from this early homogeneous cosmos.

Galaxy Evolution

Once hydrogen and helium atoms dominated the universe, the electrically neutral cosmos continued to expand. As it spread over the next few hundred million years, radiation cooling continued. As a result, the universe became virtually dark. No stars existed.

Although the initial radiation density was remarkably symmetric, as indicated by the CMB data, it was not perfect. Variations at the level of one part in a hundred thousand occurred. Estimations suggest that this tiny variation was all that was needed to seed the formation of mass structures. Because this gravitational attraction between regions that have more mass is stronger, atomic matter began clumping together into giant clouds with empty space between these clusters.

Where more matter was present, gravity continued to coalesce adjacent matter. This motion continued within the giant clouds, even as smaller clouds of atomic matter formed. Approximately 700 million years after the Big Bang, the largest clumps formed protogalaxies with the smaller clumps of atomic matter heating as they collapsed in upon themselves. This heat energized the atoms, forcing them to move more rapidly, which led to frequent, violent collisions. Eventually the heat became so intense that the electrons were stripped from the protons, creating ionized plasma, which resembled the plasma of the early universe, except that it was gathered into discrete clumps.

When temperatures of 10 million degrees were reached, the protons collided with enough momentum that they combined, forming helium through the process of nuclear fusion. During this conversion, mass was lost and converted into vast amounts of energy. The subsequent heat generated at the center of each collapsing cloud stabilized it, preventing the cloud from collapsing further. Thus, a star was born.

This process of energy formation and collapse was continuously repeated as smaller collapsing clouds of matter created billions of stars. In time, galaxies filled with stars, lighting up the universe. Each star generated heat and light as long as it had enough hydrogen to keep the fusion reactions going. Our galaxy, the Milky Way and our Sun, within the Milky Way, was formed in the same manner. In approximately 4.6 BYA our Sun was created. Today, the Sun is half-way through its life of 9–10 billion years.

For perspective, the number of galaxies in the observable universe is estimated to be about two trillion (2×10^{12}). The number of stars in the Milky Way is approximately 100 thousand million (10^{11}). The spiral structure of the Milky Way has a diameter of 106,000 light years (the distance light travels in a year). And our Sun is located about 26,000 light years from the galactic center on one of its spiral arms.

Two observations deserve mentioning before continuing: that of dark matter and dark energy. Dark matter is the additional mass needed to explain the rotation of the outer regions of galaxies. It also causes light to bend around large galaxies. In contrast, dark energy makes itself known by the finding that the most distant objects in the observable universe are receding at ever-faster rates. At present, understanding both dark matter and dark energy are active areas of research. Both operate on galactic scales and neither affects what happens on Earth.

Star Evolution

In general, a star first uses its supply of hydrogen in its core to produce helium through nuclear fusion. The next phase in its evolution depends critically on the star's initial mass. If the star's mass is small, it soon runs out of hydrogen, the core fills with the helium byproduct and fusion ceases. Without heat from the fusion process, the core collapses due to its compression from gravity. Its center heats up expelling its outer layers into nearby space. It is now called a white dwarf and becomes even smaller. Yet, because of the intense heat remaining in its core, it continues radiating as its heat slowly dissipates. Once cooled, the star eventually evolves into a cold, inert and burned-out mass. What was once a glowing star is now referred to as a black dwarf.

If the star has more mass, like our Sun, temperatures rise high enough in its outer layers for hydrogen fusion to continue. As a result, the star expands forming a red giant. In such stars, the collapse of the center creates temperatures high enough that helium fuses to form carbon. After the star depletes its helium, the core collapses. When our Sun reaches this point, present astrophysical theories suggest that it will expand until it engulfs the three nearest planets: Mercury, Venus and Earth. Carbon will then disperse throughout the nearby space before the Sun collapses in on itself and becomes a white dwarf. This will start to happen, in another few billion years.

More massive stars present a different physical dynamic. If there is more mass when they run out of helium, their cores will continue to collapse and the additional compression results in ever higher temperatures. In a sequence of volatile burns, the carbon fuses and creates other elements, such as oxygen and silicon. This order repeats as each new element is depleted. The core contracts, temperatures rise to even higher levels and the star continues to produce higher atomic-number elements. This process is structured, with different fuels being fused in different layers within the star.

Eventually, if the star is massive enough, its core reaches temperatures of 4 billion degrees Celsius and iron is created. Iron is the highest atomic-number element that can be made through the process of fusion. Once the core of a massive star has filled with iron and fusion ceases, the star collapses one last time in a massive explosion, called a supernova. Momentarily, the star shines as brightly as an entire galaxy, with most of its mass blown into outer space, forming an outward expanding ring around its remaining center.

In the explosion, enough neutrons are generated to produce the bulk of the higher atomic-number elements beyond iron in the universe. These elements need additional neutrons in their nuclei to maintain electromagnetic stability. They are not made in the previous fusion sequence. Rather, they are formed by a process of neutron capture from the exploding neutrons. Today, the Crab Nebula is an example of such a supernova. It exploded in 1051.

In time, the core of a supernova will further collapse into a dense compact mass. Depending on the star's mass, either a neutron star or a more massive black hole will be formed. A neutron star is generated when the pressures in the core of the star, become so intense that its electrons and protons merge to form neutrons. Due to its spin, it radiates electromagnetic energy in pulse rates that can be as fast as a few per milli-second or as slow as a few per second, and is referred to as a pulsar. These stars were first discovered in 1967. At the center of the Crab Nebula, for example, is a pulsar.

In 2017, two neutron stars were observed to engage and then collide. Like the explosion of a supernova, this merger produced copious amounts of heavy elements. It is now theorized that as many as half of the elements heavier than iron are produced by such collisions.

If the original star is even more massive, it can form a back hole, which is a region of space so dense that nothing can escape its gravitational field, not even light. Hence, its name. Astronomers have observed that our Milky Way has a large black hole in its center, as do most galaxies.

The first observation of black holes occurred in 2015, when the merger of two black holes was seen simultaneously by both of LIGO's (Laser Interferometer Gravitational-wave Observatory) interferometers. The LIGO "observatory" consists of two identical and widely separated interferometers situated in sparsely populated and out-of-the way places. LIGO Hanford in southeastern Washington State is in an arid shrub-steppe region crisscrossed by hundreds of layers of ancient lava flows and LIGO Livingston is 3002 km away in a vast, humid, loblolly pine forest east of Baton Rouge, Louisiana.

In 2017, LIGO detected the merger of two neutron stars. Both the black holes and the neutron star mergers were sensed in LIGO, by measuring the distortion of the Earth's surface from the passing gravitational field the stars generated. The gravitational waves in the 2017 observation were also detected simultaneously with an electromagnetic gamma-ray burst, proving that gravity waves travel at the same speed as light. The detection of such initially unimaginable creations as a neutron star or a black hole provides credibility to the current understanding of the end processes in the evolution of massive stars.

Solar System Evolution

Our Sun, like all stars, was formed from a collapsing cloud of matter about 4.6 BYA. It was likely initiated by a shock wave from a nearby supernova explosion. The debris would have produced radioactive elements, like uranium, that are found on the Earth. Analysis by astronomers suggests that our proto-Sun was probably made of material similar to that observed in a recently forming cloud in the Orion constellation. This cloud consists of 70% hydrogen, 27% helium, 1% oxygen, 0.3% carbon and small percentages of the remaining 92 natural elements.

Like other interstellar clouds, this cloud eventually flattened into a rotating pancake disk, as gravity pulled its mass inward. This force of gravity attracted matter toward its axis of rotation and the centripetal force in the off-axis region helped keep the outer regions away from the center. Thus, a bulge (our future Sun) developed around its middle. At the outer edge of the rotating cloud of debris, the outward centrifugal force eventually exceeded the inward gravitational pull and thin rings of gaseous matter broke away from the rest of the forming solar mass.

Interstellar dust, which was swept up in the process, helped cool the rings by radiation. The dust was a holdover from remnants of past solar explosions. The decrease of the outward pressure in the rings as they cooled helped facilitate their inward radial contraction from gravity. Within the rings, the largest objects had stronger gravitational fields, as they accumulated more and more mass in their orbit. This process of accretion resulted in dust condensing into chondrules, planetesimals and then planets. Not all of the rings formed planets. The region between the future Mars and Juniper, for example, remained as an asteroid belt.

The rotating matter was a mix of light and heavy elements. The heaver elements sank to form planetary cores in a cloud of lighter elements. In the regions near the Sun, the comparatively hotter and lighter elements were expelled to the outer reaches of the forming solar system. The heavier elements remained in the central core and contributed to nearby planets. Within the next 100 million years, four rocky inner planets (Mercury, Venus, Earth and Mars) formed. Closer to the outside of the emerging solar system, where it was cooler, lighter elements accreted into planets. Here, the large gas-giants Jupiter, Saturn, Uranus and Neptune developed. Approximately a million years later, the Sun exploded, producing a blast of particulate matter. This powerful particle emission, called a stellar wind, swept away the gaseous outer shell of the Sun and stripped the atmospheres from the forming rocky planets.

The inner planets remained with little atmospheric gas and only the outer planets were composed of the lighter elements.

Earth-Moon Evolution

When the solar system was 50 million years old, a proto-Earth formed in a 150 million km radius surrounding the Sun. It was not the only planetesimal to form a planet in this orbit. A smaller proto-planet, named Theia, also developed. Computer simulations using a cosmological model known as "the giant-impact hypothesis" determined that the origin of the Earth-Moon system can be explained by Theia colliding with the more massive Earth at a glancing, oblique angle. This collision jarred the early Earth slightly off its axis shattering the smaller Theia. In the process, early Earth's gravity attracted much of what had been Theia, including much of Theia's iron core. Within a few years, the bigger chunks of Theia coalesced, the remaining debris with its gravitational pull as it orbited, then only 24,000 km above the Earth's surface. This process created our Moon.

The glancing collision theory elegantly explained how the new Earth gained its anomalous axial tilt of 23°. This theory also explained the similarities between the Earth rocks and those on the Moon. The collision resulted in the angular momentum of the Earth and Moon being coupled into one spinning unit with one side of the moon always facing the Earth.

Initially, the dynamics of the collision resulted in the Earth rotating on its axis once every 5 h, while the Moon circulated the Earth in three and one half days. The surfaces of the early Earth and Moon were molten and not yet fully solidified into rock, so the circulating Moon's gravitational pull exerted huge tidal forces on the Earth's molten surface and vice versa. These surges on the Earth and Moon formed tidal bulges on their surfaces. The Earth's bulge was always in the lead (ahead), relative to the Moon's bulge, because the Earth was spinning at a faster rate than the Moon was circulating around the Earth. The leading Earth's bulge, through the force of gravity, constantly transferred angular momentum to the Moon with every orbit. The faster the Moon orbited the Earth, the farther it distanced itself from the Earth. Simultaneously, the Moon pulled back on the Earth's bulge with an equal and opposite force, making the Earth rotate on its axis more slowly with every revolution. The combined slowing of the rotations and the increasing distance between the Earth and Moon preserved the total angular momentum of the Earth-Moon system.

Currently, the Earth has slowed to one rotation a day, with the Moon circulating the Earth every 30 days. Since the Moon's formation, it has receded to 385,000 kms farther from the Earth. The process continues to the present and, according to National Aeronautics and Space Administration (NASA), every year the moon moves 3.8 cm farther from the Earth.

Earth Evolution

After the glancing Theia Earth collision, the Earth evolved into a period of self-organization and cooling. The initially well mixed and molten interior separated into layers of matter that eventually became distinct: a surface crust and upper mantle, followed by a lower mantle and an inner and outer core. Oxygen, magnesium, aluminum, silicon, calcium and iron dominated the elements that formed the outer layers. Dense, molten iron settled to the core, which then became 90% iron. Today, these six elements make up 98% of the Earth's mass, all of which resulted from earlier exploding stars.

The layers of the developing planet Earth differed structurally, because the temperature and gravitational pressure increased with depth. From the exterior, the cooler solid crust floated atop the hot upper mantle where melting occurred, enabling the crust to move independently. Although hot, the rocks in the lower mantle were solid, due to the increased pressure. In contrast, the metallic outer core formed a liquid layer, where swirling electrical currents generated the Earth's magnetic field by convection. This convection was powered by the Earth's heat and rotation. The dense hot liquid, at the boundary of the inner core, expanded and rose. As it rose, it was replaced by cooler, denser liquid that descended from above, forming convection cells. The inner core, like the lower mantle, acted as a solid despite the very high heat, owing to the increase in extreme pressure generated from the layers above.

After the collision, the Earth did not have a solid surface. Asteroids continued impacting the planet. Every collision added more thermal energy thereby heating the impacted area. Gravity induced tides, from the initially close Moon and contributed to maintaining Earth's liquid state. Earth's store of radioactive elements added heat. A growing atmosphere produced by volcanic vapors, rich in water and carbon dioxide, contributed to the heating by creating a greenhouse gas effect. This molten state prevailed for up to a hundred thousand years, as the Earth slowly spun through cold outer space, cooling by means of conduction, radiation and principally, convection.

Eventually, hot peridotite, composed of olivine and pyroxene, became Earth's earliest surface. The peridotite magma that formed was less dense than the cooling surface peridotite. The cooling peridotite solidified, deformed, broke up and sank back into the mantle, only to be reheated and partially remelted. In time, the upper mantle became mostly solid peridotite, a greenish black rock.

The hot peridotite concentrate, however, was not strong enough to support the thick, heavy, solid surface. In contrast, beneath the surface, the partially melted peridotite produced a tough rock called basalt, that was rich in silicon, calcium and aluminum. This black basalt had a density 10% lower than that of peridotite, which allowed it to rise to the surface.

Volcanic eruptions, like that of the recent Mount Saint Helens, continued to spew volcanic basalt lava and ash over the Earth's surface. The basalt lava excreted through fissures and filled pockets in the peridotite. Eventually, the entire Earth's surface was covered with basalt, which for the first time floated as a crust on the mantle.

Within a few million years after the collision with Theia, water vapor, which had been blown away during the impact, returned and became a principal component of the early atmosphere, resulting in global wide torrential rains. The water and rock interactions hastened the cooling of the basalt crust. The low lying regions of the basalt filled with water, eventually forming the first oceans. These oceans were punctuated by isolated, steaming volcanic islands protruding above the water's surface. Near 100 MY after the Theia-Earth collision, the Earth filled with water at 1–2 kms depth. Ice from incoming asteroids likely added to the formation of the ocean.

As the outer layers of basalt cooled and hardened, they formed a cover that prevented heat from escaping the partially molten mantle below. As with the previous peridotite melting into basalt, the basalt reheated from beneath, began to melt and formed a new silicon rich magma that was 10% less dense than the basalt itself. This new, lighter magma thrust its way towards the surface via volcanic action. On the surface, the magma crystalized as granite rock floating on the basalt, similar to ice floating on water. The rudiments of a future continental crust began to form.

Deep within the hot, pressurized outer mantle, the rocks were soft and flowed slowly. Within the core, the layer of cooler, denser mantle rocks descended slowly, while the hotter, lighter rocks rose toward the surface, forming giant convection cells. The dimensions of these cells occasionally reached thousands of kms across and many hundreds of kms deep. The motion of the convection cells was so slow it generally took a hundred million years or more to complete a single revolution.

In the beginning, the primordial basalt crust was relatively uniform. But, with convection cells pulling at the base of the brittle basalt crust, the crust began moving and separating into large basalt sections called plates. These plates bumped into and slid under and over one another, due to the influence of the circulating cells. A new, hot basalt crust formed where these convection cells rose upward to separate the basaltic plates. Old, colder basaltic crust simultaneously plunged into the mantle in subduction zones, areas where the convection cell descended, in a transformative process called plate tectonics.

Throughout these changes, the total basaltic magma remained constant. In contrast, subduction of the cold, wet basalt accelerated the production of granite. This magma rose to the surface during volcanic eruptions, producing a series of volcanic islands hundreds of kms long. Repetitions of this process and accretion of these island chains eventually formed continental crusts of granite floating above the original crustal basalt.

Today, the Earth is divided into seven major plates and a number of intermediate sized plates. The north–south Mid-Atlantic Mountain ridge, upon which Iceland sits, is an example of an upward welling of basalt magma derived from an ascending circulating convection cell. The South America and Africa continents, which were once connected, are now being pushed apart at a rate of several centimeters a year. This process is referred to as a rift. In contrast, the eastern motion of the Pacific Ocean plate is being pulled under the continental North American plate. This movement is a classic example of a subduction zone. A line of volcanos from the Canadian and Washington border through California, a hundred km inland from the western continental shelf, is generating new continental magma. Although this plate motion is only a few centimeters a year, over millions of years, continents are formed, deformed, and reformed into other configurations.

The current continental configuration is a result of the fracture of a supercontinent, named Pangaea that begin forming 300 million years ago (MYA) and rifting 100 MY later. Over Earth's history, at least five giant continents formed, rifted apart and fragmented into individual continents surrounded by oceans. In general, the conveyor belt of plate tectonics produces fresh ocean basalt at the ocean ridges and absorbs it under the continental shelf approximately every two hundred million years. Similarly, continents take hundreds of millions of years to break apart and reassemble.

As the surface transformed, greenhouse gases like water vapor and carbon dioxide continued to emerge from the mantle via volcanoes and cracks in the Earth's surface to form the first atmosphere. The atmosphere and crust were further enhanced by water and gases deposited by comets and asteroids. Between 4.0 and 3.8 BYA the Earth and Moon were hit by massive bombardments of asteroids, known as the Late Heavy Bombardment. In these many processes, the corrosive impact of water and wind helped stabilize the Earth's temperature. Tectonics builds mountains of granite on the continents, erosion grinds them down. Carbon dioxide dissolved in rainfall forms carbonic acid which, in turn, dissolves more rock. Carbon, a byproduct of these reactions, is then removed and deposited in the oceans, where it becomes carbonate rocks, such as limestone, which are buried in the mantle for eons. Eventually, some of the removed carbon dioxide is returned to the atmosphere though volcanism in tectonic subduction zones of the mantle. A balance between the two mechanisms of emission and erosion occurs. Higher temperatures from an increase in carbon dioxide from volcanism generates greater rainfall, which, in turn, accelerates erosion. This sweeps more carbon back into the mantle, leading to a reduction of carbon dioxide and to a cooling the Earth.

Chapter 3
Life

Primitive Life and Snowball Earth

Prokaryotes

Within the first 50 million years of Earth's formation the required elements for life existed. Volcanic emissions containing water, carbon dioxide, nitrogen, potassium and sulfur easily reacted with rocks and sea water to create almost all of life's molecules. In the mid-oceanic ridges, hydrothermal vents where the Earth's tectonic plates drifted apart, hot water loaded with dissolved minerals and volcanic gases, necessary for life, arose. There, life forms perhaps combined hydrogen with sulfur to form hydrogen sulfide for its energy needs. Sheltered from the Late Heavy Bombardment, it was an ideal location for initiating life. An alternative theory postulates, however, that organic molecules found on asteroids may have seeded early life.

Over eons, as Darwin's Theory of Natural Evolution explains, chemical reactions likely evolved through the process of natural selection to build long chains of amino acids (the molecules of proteins) and nucleotides (the molecules of nucleic acid). Proteins are the precursors needed for cell maintenance and metabolism; nucleic acid is required for coding reproduction. It is theorized that these basic units of life emerged as a self-contained cell where chemical reactions occurred. The cell, enclosed by a leaky membrane made of lipids (which easily formed bubbles), protected the reactions, let in sources of energy and permitted waste products to leave.

Researchers believe that these earliest cells may have been based on the autocatalytic molecule ribonucleic acid (RNA) before the more complex deoxyribonucleic (DNA) formed. RNA enabled the synthesis of proteins, both by carrying genetic coding and by catalyzing the formation of protein. The resulting cell life likely emerged as an autocatalytic chemical reaction that took energy and matter from its surroundings in order to replicate, grow and propagate itself.

© The Author(s), under exclusive license to Springer Nature Singapore Pte Ltd. 2024   15
T. Sanford, *A Whirlwind History of the Universe and Mankind*,
https://doi.org/10.1007/978-981-97-2674-5_3

As a cell grew by feeding on surrounding nucleotides, developing fatty molecules of the cell wall, it likely elongated and split into two spheres. Some cells reproduced faster than others based on the nature of their molecules. At this stage, the autonomous cells metabolized, replicated and the evolution of life began. The details just described are possibilities. The actual origin of life's beginning is currently an active area of research.

The first living cells were likely similar to today's simple, single cell microorganisms called prokaryotes, such as archaea or bacteria. Archaea are heat loving and formed in deep-sea vents; bacteria are formed everywhere else. Archaea use sulfur for their source of energy, whereas bacteria need to eat organic (carbon) compounds. Geochemical reactions between rocks and water generated the hydrogen sulfide and methane that the prokaryotes use.

By 3.5 BYA, the early atmosphere contained an abundance of gases, water vapor, carbon dioxide and methane but no oxygen. Some bacteria, as is true today, were able to use the methane for food. The first fossil evidence of life in the form of stromatolites was found in rock dating to this time. These structures were composed of bacterial mats formed in shallow sea water. They employed an early form of photosynthesis that did not produce oxygen as a by-product. The energy generated came from the sunlight's interaction with hydrogen sulfide or iron that dissolved in the ocean.

This transformation of energy and matter from the outside environment was in contrast to how the evolving universe formed galaxies into stars and dissipated energy. Within a galaxy, matter was self-contained and grew by its gravitational attraction. Energy was generated internally, fusing some of its protons into more complex nuclei. During this process some of its mass was turned into energy.

Great Oxidation Event and First Snowball Earth

By 2.4 BYA, cyanobacteria made the evolutionary steps that produced oxygen. By extracting energy from the sunlight through the process of photosynthesis, using water and carbon dioxide as food in an oxygen emitting system, these cells were able to produce carbohydrates and oxygen. As the cyanobacteria consumed carbon dioxide, the resulting free oxygen broke down atmospheric methane. Carbon dioxide and methane are both critical greenhouse gases. Within a million years, oxygen levels quickly rose from less than 0.001% to greater than 20% of the atmosphere. This change was known as the Great Oxidation Event.

As the oxygen levels rose and greenhouse gasses diminished, the Earth cooled. Eventually, the continents were covered in ice and the oceans froze. The surface of the Earth turned white, thereby reflecting more sunlight, cooling the Earth further and formed the first snowball Earth. In this positive feedback loop, life forms retreated to the oceans near regions of warm volcanic activity. Later, in a reversal of these trends, ice blocked photosynthesis and greenhouse gases from volcanos built-up under the ice and eventually broke through. The result was the Makganyene glaciation, which lasted from 2.35 to 2.22 BYA. Oxygen levels plummeted and finally stabilized at 1–2% of the atmosphere for the next billion years when the climate remained warm.

During this long epoch, an over abundance of sulfur using bacteria competed with the new, oxygen releasing types. This first life used H_2S for photosynthesis in contrast with those that today used H_2O. Both competed for nutrients and space. Although the splitting of water for photosynthesis was a far more difficult process, the use of water in oxygen generators, rather than sulfur photosynthesizes, became the norm. The introduction of free oxygen molecules radically transformed the biosphere. Once in the high atmosphere, the oxygen formed ozone, which in turn began to shield the Earth's surface from the sun's bio-damaging, ultraviolet radiation. On Earth's surface, new minerals were formed.

Prior to 2.4 BYA, the oceans were filled with dissolved iron and low in sulfur. This configuration was able to be maintained because water lacked oxidants, which cause iron oxides to precipitate. Post 2.4 BYA, the oxygenating water resulted in the formation of pyrite and other iron sulfide minerals. With the increase in oxygen, some of the iron formed iron oxides directly in shallow waters or indirectly with sulfur bearing minerals in the oceans that had eroded from the oxidized, weathered land. As a result, between 2.4 and 1.8 BYA, enormous amounts of iron minerals were deposited in layer upon red-and-black layer on the ocean floor. This precipitation is known as BIF (banded iron formations). Today, the BIFs constitute 90% of the Earth's known iron reserves.

Eukaryotes

Eukaryotes were the first cells to exploit the chemical energy of oxygen systematically through the process of respiration. Respiration uses oxygen's enormous chemical energy to convert the energy stored in carbohydrates into water and carbon dioxide as waste products. Respiration extracts more energy from organic molecules than the earlier, non-oxygenic way, of digesting food molecules.

Genetic evidence indicates that the first eukaryotes evolved following the oxygen transformation, approximately 1.8 BYA. Eukaryotes were the third domain of life forms after the archaea and bacteria formed. As oxygen became more readily available, eukaryotes became the organism of the future. Eukaryotes removed much of the atmospheric oxygen generated by the cyanobacteria and together with the competing sulfur-using bacteria, helped explain why the climate remained relatively stable after the Snowball Event.

Eukaryotes were formed in the process of endosymbiosis, whereby one cell absorbed another whole as opposed to digesting it. Eukaryotes, an independent bacteria (with its own DNA), began by living inside an archaea cell with its independent DNA. Other bacteria followed, such as mitochondria, an oxygen respiring bacterium, or chloroplasts, a cyanobacterium that conducts photosynthesis.

Eukaryotes reproduction marked a decisive change in life's evolution. Eukaryotes, with their independent organelles, grew to sizes 10–1000 times that of prokaryotes with their DNA enclosed in a membrane that constituted a well defined nucleus. Prokaryotes pass their genes on by splitting in two with little change in their genetic structure. In contrast, eukaryotes developed a new way of reproduction called sexual reproduction. With sexual reproduction, half the genes of one parent cell were combined with half from the other. This process resulted in a greater variation in genetic material than that achieved by simple mutations. With the increased change in offspring, evolution accelerated. Eukaryotes provided an example of evolution through symbiotic collaboration rather than pure competition for resources. They became the precursors of all big cells whether plants or animals.

Three Additional Snowball-Hothouse Cycles

Although the Earth was remarkably stable for more than a billion years after the Great Oxidation Event, its biosphere had changed markedly. By 850 MYA, the rising oxygenated ocean shores exploded with microorganisms, including algae and sulfur-processing bacteria. Most of the Earth's continental masses were connected near the equator in a lifeless, rusty-red supercontinent named Rodinia. This situation would not last. By 750 MYA, tectonic shifts under Rodinia were underway. These rifts split Rodinia in half. New coastlines and inland seas formed. Coastal erosion and weathering, caused by rain mixed with CO_2, formed carbonic acid and contributed mineral nutrients to the photosynthetic algae. Microbial life mushroomed, thereby drawing down carbon dioxide and increased atmospheric oxygen. Evaporation from the inland seas caused more rainfall, which in turn, weathered the surface rock and further reduced the carbon dioxide. The climate cooled and ice caps formed at the poles. This increased albedo of the ice reflected more sunlight and the cycle addition-ally cooled the surface until ice encased the Earth again. As in the Great Oxidation Event, constant tectonic shifts eventually allowed volcanic carbon dioxide to crack through the ice. This evolution finally collapsed the cooling cycle and Earth slowly warmed.

Over the next few million years, Earth repeated two more periods of cooling and warming. In total, the above Sturtian glaciation peaked 720 MYA, the Mari-noan glaciation peaked 650 MYA and the Gaskier's glaciation 580 MYA. Through each of these events, oxygen levels rose to approximately 6, 7 and 10%, respec-tively. These severe climatic cycles never reappeared again. They, however, increased the evolutionary pace of life, driving many organisms to extinction and forcing the development of new organisms.

From 750 to 600 MYA single celled, photosynthetic plants appeared on damp land surfaces although soils did not yet exist. Multicellular plants, such as kelp and algae, evolved on seashores. The oceans filled with life-like amoeba and paramecia. A unique feature of this era were the half-plant, half-animals such as the multicellular volvox.

Complex Life and Mass Extinctions

From the remaining years to the present, the continents continued to move, break apart and reconfigure. By 300 MYA, the remains of Rodinia began to reshape into the supercontinent Pangaea. By 250 MYA, Pangaea fused into a single land mass and by 200 MYA it too began rifting into the present continental configuration. These changes influenced climate fluctuations from cold to hot and vice versa. Oxygen levels also increased, decreased and rebounded. Sea levels fell and rose repeatably, often by hundreds of meters, to impact coast lines. Throughout these transformations, rocks and life morphed. In contrast to the microbes, new sea and land animals and plants were particularly sensitive to the environmental and climatic changes.

Fossil discoveries allowed geologists to divide the last 635 MY into thirteen clearly defined periods of life, separated by mass extinctions. The extinctions were the result of tectonic plate motion, asteroid bombardment and changes in oxygen and carbon dioxide levels due to volcanic eruptions. The transition from one period to another was marked in the fossil records by the loss of some fauna at the beginning of each period, and the appearance of new fauna. The new fauna appeared to be related to environmentally stressful episodes when evolutionary innovation was connected to new ecological opportunities not yet dominated by former species.

The thirteen periods, generally defined by their fauna were the: Ediacaran (635–541 MYA), Cambrian (541–488 MYA), Ordovician (488–444 MYA), Silurian (444–416 MYA), Devonian (416–359), Carboniferous (359–299 MYA), Permian (299–252 MYA), Triassic (252–200 MYA), Jurassic (200–145 MYA), Cretaceous (145–66 MYA), Paleogene (66–23 MYA), Neogene (23–2.6 MYA) and Quaternary (2.6–0 MYA). These periods were further combined into three eras: Paleozoic (541–252 MYA), Mesozoic (252–66 MYA) and Cenozoic (66–0 MYA).

In five of the extinctions (at the end of the Ordovician, Devonian, Permian, Triassic and Cretaceous periods), over 50% of the species were lost. During the largest extinction, the transition between the Permian and Triassic periods 252 MYA, 90% of all species died. This particular extinction marked the end of the old Paleozoic Era, which began 541 MYA. The following Mesozoic Era lasted until the massive asteroid impact of 66 MYA that practically ended life at the end of the Cretaceous Period. The following Cenozoic Era to the present completes the three eras of the Phanerozoic Eon (541–0 MYA).

Ediacaran Period (635–541 MYA)

The first fossil evidence of a multi-celled, animal dominated ecosystem closely followed the Gaskiers glaciation. Soft bodied animals, like jellyfish and worms, left fossilized impressions in the ocean rocks, during the Ediacaran Period. It was during this period that animal locomotion and feeding first appeared in ocean sediment. These impressions implied Ediacaran's developed non-mineralized skeletal structures, muscles, nerves and a sensory system in order to find food, mates and avoid predators. Eventually, bilateral symmetry evolved and locomotion improved. In tube-like bodies, these animals developed a well defined front and rear end with a digestive system in between.

Cambrian Period (541–488 MYA)

By 532 MYA, the Ediacaran's suddenly disappeared and new animals emerged. These new arrivals evolved protective, durable hard parts that were easily identifiable in the fossil record. This change is commonly known as the Cambrian Explosion. The iconic arthropod (invertebrate) fossils of the period were trilobites and clam-like brachiopods. Modern day horseshoe crabs are among their descendants. The arthropods had modular parts with reductant morphologies that were used to facilitate different evolutionary functions. These new creatures were also transforming the Earth's surface as their calcium carbonate skeletons and shells accumulated and built thick layers of chalk on the ocean bottoms.

Ordovician Period (488–444 MYA)

This period was the beginning of blooming plant life. Near the beginning of the Ordovician Period, early aquatic, green algae found land and became the first leafless, vascular (those with roots and stems) plants. Now anchored, these centimeter high plants were similar to low spreading moss. Coral reefs proliferated during the period, quickly increasing in size, type and distribution, until the Devonian Period. The coral, in turn supported a vast diversity of plants within its tissues. Combined with the survivors of the Cambrian Period, the second part of a two-stage animal life occurred on the Earth. In both periods, rising oxygen levels was the driver.

Silurian Period (444–416 MYA)

The oldest fossil records of land animals occurred during the Silurian Period. They all were small anthropoids. A lineage of proto-scorpions with water gills emerged from the lakes and freshwater swamps by 430 MYA. These were followed by crawling millipedes around 420 MYA and flying insects by 410 MYA.

Devonian Period (416–359 MYA)

Like the animal multiplication, during the Cambrian Period, the Devonian Period produced a rapid diversification of land-based plants. By 385 MYA, plants with leaves appeared. Competition for light among low growing plants resulted in the development of tree trunks for height. By the end of this period, trees grew to more than eight meters in height with roots extending downward a meter. Woody trees covered nearly every landscape, causing atmospheric oxygen to rise. At the beginning of this period, oxygen measured from a low of 5–10% to levels as high as 30–35% by the middle of the Carboniferous Period.

As plant life changed, so did the planet. Nutrient rich soils formed as the primitive plants died and decayed. Landforms were transformed by chemical erosion as wind and water eroded underlying rocks. The increased weathering removed carbon dioxide from the atmosphere, reversing the greenhouse effect. The planet was then plunged into a long running ice age by the start of the next period. Prior to the cooling, however, the atmosphere had adequate levels of carbon dioxide and thus remained warm throughout the period.

Although the first fish developed late in the Cambrian Period, the Devonian Period was also known as the Age of Fish. The first fish had small mouths and no jaws, which limited their feeding. Later, the seas were filled with competing, predacious fish similar to today's sharks, lampreys, bony fish and the now extinct placoderms (first jawed fish). Some grew to more than 5 m, while others developed heads larger than a soccer ball. There was motivation to get big, get armored or to crawl out of the water to survive.

Until 365 MYA, most animal life had evolved in the ocean. The exit from the water was driven by the diminishing shallow seas and shorelines among the continental pieces. By 300 MYA, parts of land mass were converging into the supercontinent Pangaea. Competition for food as the seas shrank became fierce.

The transition from aquatic animals, like fish, to creatures capable of walking on land, however, began before the seas diminished. In 2004, a 6 years search was conducted in rocks formed in streams 375 MYA, by Neil Shubin and his team. They discovered fossilized proof of the evolutionary transition from fish to land walking animals. This go-between was named Tiktaalik. Shubin's discovery was a testament to his teams understanding of the time when such an animal could exist and their grasp of environmental conditions where such creatures might be found.

Fish have conical heads and no necks. Tiktaalik had a flat, crocodile-like head, with eyes on top and a neck. An other evolutionary feature that Tiktaalik possessed was four bony fins that were the forerunners of limbs. The bones of the front two fins correspond to those in the upper arm, forearm and wrist of early land animals. The rear fins evolved into legs. But, like a fish, Tiktaalik had scales. Importantly, beyond illustrating the evolutionary transition from fish to amphibian, Tiktaalik was the precursor to all tetrapods, including ourselves. Our own DNA supports this unique ancient lineage.

Carboniferous Period (359–299 MYA)

At this time, Pangaea was forming and the early Atlantic Ocean disappeared. Enormous mountain ranges ran north and south as North America squeezed together with Europe, and South America fused with Africa. The new mountains captured dense clouds and a wet climate reigned over the bulk of the Earth. Large flood plains formed on either side of the mountains. The evolving trees settled into vast swamps and connected uplifted and drier lands.

The Carboniferous forests substantially increased the rates of photosynthesis, yielding greater oxygen production. Using cellulose and lignin for skeletal support, trees grew tall but fell easily as they lacked a deep root system. Unlike today, fallen trees did not immediately decompose. Organisms to breakdown the lignin had not yet developed. These grand forests were thus largely buried beneath eroding soils along with the carbon that they had drawn from the atmosphere. This meant that the buried carbon could no longer react with oxygen to form carbon dioxide. With less carbon dioxide, oxygen levels rose. Extraordinary levels were reached around 330–260 MYA, perhaps as high as 35% of the atmosphere. Eventually, under pressure, the buried carbon fossilized, and formed 90% of the Earth's coal deposits.

The elevated oxygen levels reached in the Carboniferous Period allowed animals to grow to sensational sizes. In general, the increased size helped to protect against predation. Scorpions grew to lengths of a meter and weights of 20 kg. In insects, with more oxygen available, it diffused through their respiratory systems, permitting large sizes to prevail. The wingspans of dragon flies reached a meter with body widths of centimeters and lengths of a third of a meter.

Amphibians were also affected. Between 340 to 330 MYA new amphibian species that could breathe air and support their heavy bodies with limbs crawled and then scampered across the land. Initially their legs splayed to the sides of their body like a lizard's. This configuration made running and simultaneously breathing impossible. They eventually evolved enormous species of their own. Before the period ended more than 320 MYA, three independent groups of reptiles developed from the amphibians: aynapsids, which became turtles; diapsids, which gave rise to dinosaurs, crocodiles, lizards and snakes; and synapsids, which included the ancestors of mammals. Between the fauna and flora, the Carboniferous Period was an era of giants.

Permian Period (299–252 MYA)

The Permian Period may be considered the first age of mammals. In the Permian, the synapsids further evolved and diversified into therapsids, which, in turn, gave birth to mammals. Synapsids included more than 70% of the land vertebrae by the start of the Permian Period. They possessed varied appetites being both carnivores and fish eaters as well as the first large herbivores. At the peak of the oxygen levels they became the biggest of all land vertebrae. Both carnivores and herbivores reached lengths of nearly 5 m. Their legs, rather than being splayed to the sides, were positioned under the burden of their bodies creating an upright stance. This modification significantly decreased the compression on the lungs, during movement, allowing for improved breathing throughout rapid motion.

Both predator and prey developed large sails on their backs. Aside from making them appear larger and more formidable, these sails functioned as sun-ray catchers to warm their bulky bodies prior to a chase or flight. As oxygen levels dropped in the later part of the Permian, the size of animals diminished. In general, the fossil record shows a strong correlation between animal size and oxygen levels throughout the Permian and into the next Triassic Period.

Triassic Period (252–200 MYA)

The most deadly of all mass extinctions occurred between the Permian and Triassic Periods when 90% of all species were extinguished. At the start of the Triassic Period, only a fraction of the land surface was inhabitable and the world was largely devoid of life. The continents, at this time, were fully merged into the supercontinent of Pangaea. The lower oxygen levels along the shoreline were the equivalent to living in the thin air of Mt. Rainier's 4395 km peak. Life on land was harsh. Oxygen fell to its lowest level in the late Triassic Period and was accompanied by hot temperatures, due to high carbon dioxide levels.

Like the Cambrian explosion of animals, new animals developed during the Triassic Period, with many new body types that were defined by competition. Previously, the inhabitable land was filled with animals that had adapted to the high oxygen levels. At this point, only those adapting to the lower oxygen levels were able to advance into the Jurassic Period. One example of the more unique morphologies was that of the flying pterosaur, which evolved in the late Triassic Period.

Animals needing oxygen for high levels of activity, developed respiratory systems that were better able to adapt to the lower oxygen levels. Among these were the dinosaurs that evolved during the later third of the period. The dinosaurs survived this extinction better than most other vertebrae because their air-sac lungs gave them the advantage of a superior respiration system. Initially, the majority of their shapes were those of the carnivorous bipedal saurischians. The bipedalism of the saurischian dinosaurs allowed for simultaneous breathing and motion. Eventually, dinosaurs assumed three generic shapes: from bipeds to quadrupeds with short necks and to quadrupeds with long necks. Towards the end of the Triassic Period, herbivorous quadrupedal saurischians (sauropods) emerged. Early on, the size of the dinosaurs was modest, 1–3 m. But, by the period's end, the larger sizes assumed dominance.

Jurassic Period (200–145 MYA)

After the Triassic-Jurassic extinction, like all previous extinctions, there were fresh opportunities for life forms to fill the Earth's empty niches. The efficiently breathing dinosaurs that managed to survive in the low-oxygen environment of 10–12% at the conclusion of the Triassic Period were poised to fill this gap. Saurischian bipeds, along with long-necked quadrupeds, prospered. Oxygen levels slowly rose to 15–20% toward in the end of the Jurassic Period.

With the rise in oxygen levels, huge versions of the earlier bipedal carnivore Allosaurus and Tyrannosaurus Rex forms dominated. The ornithischians, which diverged from the saurischians in the late Triassic Period, significantly increased in size. The best example of this was the heavily armored stegosaurs. By the end, and through out the Cretaceous Period, these dinosaurs were the most efficient and largest land animals to ever exist.

As the dinosaurs roamed, other changes were slowly taking place. As their hip joints attest, today's birds evolved from some of these efficient, air breathing creatures around 150 MYA. The correlation of increased size with increased oxygen level also played a role among some other land animals. Reptiles were but one example; in the seas were others. As long as the dinosaurs were present, however, mammals remained small and close to the ground.

Cretaceous Period (145–66 MYA)

Prior to 150 MYA, conifers and ferns (gymnosperms) dominated the plant world and provided fodder for the fauna. The sauropods, for example, fed on their pine needles. By the early part of the Cretaceous Period, a new type of plant emerged, a flowering one with broad leaves. These plants, the angiosperms (meaning hidden seeds in Greek), with their flowers and fruits, competed with and ultimately dominated the earlier flora. By the end of the Cretaceous Period, angiosperms comprised 90% of all vegetation. They are the plants that make up today's woodlands and forests.

The appearance of these new plants required that fauna's teeth have different biting surfaces. The teeth, lips and tongues designed for slicing pine needles from trees soon become obsolete. Accordingly, as the plants changed, so did the herbivores and carnivores. Although the ornithischians lacked the effective respiratory mechanism of the saurischians, when it came to food acquisition, they were superior. Ornithischians possessed bigger heads with stronger jaws and more powerful teeth and soon dominated the later Cretaceous Period. By the late Cretaceous Period, few sauropods remained.

Throughout the Jurassic and Cretaceous Periods, the sea rose and fell significantly. During the mid-to-late Cretaceous Period through the early Paleogene Period, the sea rose high enough that it divided North America into two distinct mini-continents. The gap between the continents filled with a shallow sea from north to south called the Western Interior Sea. At one point, the sea level increased to 300 m higher than todays level. Globally, the atmosphere was hot and humid over much of the Earth during the last three periods, with the exception of the final 5 million years of the Cretaceous Period.

During the Jurassic and Cretaceous Periods, oil and natural gas was formed. At that time, a hot Tethys Ocean circulated around the Earth just north of the equator between North and South America, between Africa and Europe, and through Asia. There was no north–south global oceanic conveyor belt circulating water through the ocean's depths to cool the Tethys. The hot ocean surfaces generated enormous quantities of phytoplankton and zooplankton, which were eaten by fish, forming the foundation of the ocean's food web. As the plankton died, they drifted down and together with the slowly sinking grains of minerals washed in by rivers or blown in by wind from the continents, formed the Tethys Ocean floor.

Warm water holds less oxygen than cold water and the resulting ocean floor became an oxygen-starved dead zone where bacteria were unable to breakdown the organic matter. The floor turned into a thick mire of organic-rich mud that was unable to decompose. Time morphed that floor into major deposits of black shale rock. The shale became the starter material for forming crude oil and natural gas. Today, these deposits are found throughout the Persian Gulf, the North Sea, western Siberia, the Gulf of Mexico and Venezuela.

Cretaceous-Paleocene Mass Extinction (66 MYA)

By 67 MYA, the Earth was being preconditioned for a mass extinction prior to its 66 MYA asteroid impact. Between 66.5 and 65.5 MYA, massive lava flowing from the Deccan Traps in India were emitting carbon dioxide and methane. These flows produced a deadly mechanism similar to earlier greenhouse extinctions. The oceans heated into stagnancy, anoxia occurred and many species died. Emissions from the Deccan Traps set the stage for a mass extinction. The asteroid provided the *coup de gras*.

This Cretaceous-Paleocene (K-Pg) mass extinction of 66 MYA ended the long reign of the non-avian dinosaurs along with 75% of all species. It was the clearest example of a mass extinction due to an asteroid impact. The asteroid was estimated to be 10–15 km in diameter and produced Chicxulab, a 150–180 km diameter crater just north of the Yucatan peninsula.

Proof of the asteroid impact and extinction was supported by comparing the Chicxulab Crater's age to that of the of the global ejecta layer of iridium (rarely found on Earth, but often found in asteroids). This layer was inter-mixed with shocked quartz (the result of intense pressure generated by the impact) found in strata around the Earth. Estimates suggested that atmospheric dust produced months long blackouts hampering photosynthesis. As the land rapidly cooled, a decrease in precipitation by 90% lasted for months. Plants died along with the animals dependent on eating them. The impact also produced bits of super-heated rocks that rained from the sky, igniting fires world-wide.

The Cenozoic Era (66–0 MYA), which follows this K-Pg extinction, is divided into three periods: the Paleogene (66–23 MYA), Neogene (23–2.6) and Quaternary (2.6–0 MYA). The Paleogene Period is further divided into three Epochs: Paleocene (66–56 MYA), Eocene (56–34 MYA) and Oligocene (34–23 MYA). The Neogene Period is partitioned into the Miocene (23–5.3 MYA) and Pliocene (5.3–2.6 MYA) Epochs. And lastly, the Quaternary Period is split into the Pleistocene (2.6 MYA–11,700 YA) and Holocene (11,700–0 YA) Epochs. As much is known about these divisions, this epoch time structure is now used to continue the Earth's history.

Paleocene Epoch (66–56 MYA)

At the beginning of the Paleocene Epoch, the configuration of the continents resembled that of today's but were closer together. The land bridge between North America and Eurasia remained above sea level into the next epoch, allowing animals to migrate between the two continents. Forests returned but were denser owing to the absence of voracious large dinosaurs. By 7 MY after the K-Pg extinction, the temperature stabilized to 20 °C or more, with polar sea temperatures differing from equatorial temperatures by only 10–15 °C.

Prior to the mass extinction, mammals were held in check by the dinosaurs through predation not competition. By 300,000 years after the extinction and without dinosaurs predators, mammals increased in size and diversity. Early on, they resembled small rodents. Later, dog-sized to bison-sized animals evolved. The early herbivores were leaf or fruit eaters as grasses had not proliferated. Their teeth and jaw bones allowed them to chew leaves and fruit efficiently late in the epoch. Ultimately, separation of mammalian ear and jaw bones permitted the skull to accommodate a larger brain that would contribute to future evolution.

Similar to mammals, the birds also made a transition. Initially, they competed with mammals for resources. The flightless but fearsome diatryma, for example, stood 2.4 m tall and was a dominate predator equipped with a powerful slicing beak and talons. It had estimated running speed of greater than 100 km/h.

The length of the Paleocene Epoch was shortened by sudden, global, volcanic activity and the rise, once again, of greenhouse carbon dioxide and increased temperatures. This, in turn, melted the frozen methane hydrates along the continental slopes and exacerbated the temperature increase. This sudden rise in global temperature (one of the most rapid in history) is referred to as the Paleocene-Eocene Thermal Event (PETM). This event precipitated another mass extinction and set the stage for the next epoch.

Eocene Epoch (56–34 MYA)

Following the PETM, the old plants revived but the mammals did not. As with past extinctions, new morphologies appeared with ever-greater complexity. By the end of the early Eocene Epoch, many new varieties of mammals appeared and doubled in number such that a hundred or so of today's species came into existence.

Primates (anthropoids) evolved at this time. Initially they were small climbing creatures with fore and aft limbs that could grasp—the forerunners of hands and feet. Variants of modern, hoofed herbivores emerged (ungulates) as either odd toed (horses and rhinos) or even toed (sheep, goats, pigs, cattle, bison and camels). Small rodents and carnivores evolved alongside elephants, while crocodiles swam in the northern polar regions.

In the early Eocene Epoch, forests covered a greater portion of the Earth than today. The trees expelled water vapor into the atmosphere through the process of transpiration. Much of the water returned to Earth in the form of rain. This process was especially true in tropical regions, where high concentrations of water vapor dampened a broad swath of the Earth. Aided by higher concentrations of carbon dioxide and by water vapor from the forests, the greenhouse effect continued the Earth's warming process.

By the mid-Eocene Epoch, however, the Earth began to cool. The cooling was due, in part, to tectonic plate movements in the south polar region. Initially, Antarctica was part of Gondwanaland, which consisted of the modern continents South America, Australia and India. This land mass remained temperate because its coasts were warmed by higher temperature water from the lower latitudes. As South America and Australia broke from Antarctica, the Atlantic Ocean widened. Simultaneously, the Drake Passage between the bottom tip of South America and Antarctica opened. This change allowed for a circumpolar current to encircle Antarctica.

Warm waters from the South Pacific and the Atlantic Oceans, which curled northward, were trapped in this circumpolar flow. These waters cooled as they circled Antarctica because of less solar warming. Antarctica consequently provided a platform for the build-up of large ice sheets at the South Pole. At the same time, the northern Polar Ocean was trapped by the northern hemisphere's continents, which isolated the region again from warm equatorial currents. Eventually, the Indian plate merged with the Asian plate, forcing the rise of the Himalayas. Weathering accelerated and the rate at which atmospheric carbon was sequestered in the Earth's crust increased, creating another loop for global cooling.

Oligocene Epoch (34–23 MYA)

By 34 MYA, a massive glacier began forming over Antarctica. This glaciation was related to global climate change. The worldwide climate, once uniform and stable, had transitioned to one with significant seasonality. Climate change, glacial expansion and associated extinctions characterized the transition between the Eocene and Oligocene Epochs.

By 32 MYA, two million years after the expanding Antarctic glaciers reached the sea, the slowly declining carbon dioxide level continued to drop precipitously. Cold, dense polar waters sank, creating a layer that exists to this day. Extinction of some marine life and many changes in flora followed from the consequent aridity.

Wet tropical and subtropical forests covered much of Eurasia and North America prior to the extinction. As these forests shrank, more climatic cooling and drying occurred. Furthermore, the cooling oceans generated less moisture via evaporation creating cycles of continuous climate change. Throughout the Oligocene Epoch, grassy regions and dry woodlands replaced the tropical forests on every continent. The Earth was evolving to assume the shape of its modern environmental appearance.

Many Eocene fauna disappeared, during the Oligocene Epoch, as more contemporary mammals appeared. The largest land mammal of all time in the rhinoceros family, which stood 5.5-m tall at the shoulder, was one example of a dying species. These mega-mammals, however, were replaced by many of the fauna groups living today. Mammals became increasingly modern in appearance. Many of the new species evolved molar teeth designed to grind the coarse vegetation of the open range. Although animals classified as apes were present in the Eocene Epoch, they were quite small in comparison with their future descendants.

Miocene Epoch (23–5.3 MYA)

During the Miocene Epoch, the decline in temperature continued as the seas surrounding Antarctica cooled. This deterioration resulted in drier conditions. Dense forests and open woodlands continued yielding to encroaching grasses, weeds and herbs that adapted to the seasonal droughts. The Miocene Epoch became the Age of Grasses. The spreading grasses stimulated the reproduction and dispersion of small animals; mice and rats devoured all grasses, weeds and herbs. Snakes then arrived to eat the small animals. Many groups of herbivores also proliferated because of the spreading grassy woodlands.

During the Miocene Epoch, C3 grasses, which use a three-carbon compound, dominated the grasslands. These grasses required a cool, moist growing season. Towards the end of the Miocene Epoch, these grasses were replaced, in part, by silica-rich C4 grasses, which use a four-carbon compound that thrived, during warm, moist growing seasons. Small bodies of silica contained within these grasses caused the teeth of the animals that graze on them to wear down rapidly. Thus, only those grazing herbivores with tall molars could take advantage of the C4 grasses. These grasses, responding to warmer and rainier growing seasons, spread across much of the Earth. In those regions herbivores lacking long teeth became extinct.

Along with the herbivores of the Miocene Epoch, the carnivorous mammals assumed current forms.

The great apes, which lived in the old world, diversified significantly, during the Miocene Epoch. Evidence of these early apes was found in fossils dating from 20 MYA in Africa. Because the fossils indicated significant evolutionary development, during this epoch, the Miocene Epoch is also referred to as the Age of Apes. Shortly after Africa collided with Eurasia as Gondwanaland broke apart, bands of apes, many now extinct, spread from Africa to Eurasia 15–16 MYA. This migration occurred with other isolated groups of mammals, including giraffes and elephants.

Pliocene Epoch (5.3–2.6 MYA)

In the early Pliocene Epoch, warm climates spread to higher latitudes and sea levels rose. The Pliocene's warm climate ended with the start of the modern ice-age (approximately 3.2 MYA). The Northern Hemisphere cooled to the point that vast ice sheets formed and sea water was locked in glaciers on land. These sheets, centered over North America, Greenland and Scandinavia, expanded and contracted over long periods of time. In many regions, climates have continued to remain cooler and drier. An example is the aridity that led to the expansion of Africa's Sahara Desert. Before the Ice Age, tropical forests extended to the present southern limit of the Sahara.

This cooling resulted from the closing of the gap between the continents of North and South America through the Isthmus of Panama (3.5–3 MYA). Prior to the closing, mixing of the Pacific and Atlantic Oceans restrained the salinity of the Atlantic. The more buoyant Atlantic waters, flowing northward into the Arctic Ocean, had kept the polar region relatively warm. With the greater salinity, the denser Atlantic Ocean sank prior to entering the Arctic Ocean isolating the Arctic from the warming of the circulating ocean. The increased salinity of the Atlantic Ocean was a result of the great quantities of surface water evaporated due to the dry trade winds blowing westward from the Sahara Desert across the Atlantic Ocean and being deposited as rain on the Pacific side, which lowered the salinity of the Pacific, as it was unable to mix through the Isthmus.

Pleistocene Epoch (2.6 MYA-11,700 YA)

The variation in the depth and length of the ice sheets became significant at the start of the Pleistocene Epoch 2.6 MYA. The expansions and contractions of the ice sheets were the result of the polar regions developing greater sensitivity to the slight cyclical changes in the Sun's radiation that heated the surface of the Earth, as it revolved around the Sun. These changes are referred to as the Milankovitch Cycles after the scientist who discovered them in the early 1940s. In general, the glaciation in the Northern Hemisphere was twice that of the Southern Hemisphere. Revolving ocean currents around Antarctica limited the southern ice sheets from spreading too far from land.

The fastest of these cycles, the precession of the Earth around its 23.5° tilt axis (which gives the Earth its seasons) has a wobble of 26,000 years and is referred to as the precession cycle. The tilt itself cycles through 24.5°–22.1° every 41,000 years and is called the obliquity cycle. The orbit of the Earth around the Sun is not perfectly circular; it is slightly elliptical. This very tiny variation from circularity, due to the gravitational pull of Jupiter, is referred to as the eccentricity cycle. This cycle has a period of 100,000 years. Prior to 0.9 MYA the obliquity cycle of 41,000 years dominated the measured periodicity of the ice sheets. From 0.9 MYA to the present, the 100,000 year eccentricity cycle has dominated.

When the first Homo sapiens appeared, approximately 200,000 years ago (YA), the Earth's climate was temperate; 5000 years later the climate deteriorated and the Earth entered a prolonged ice age that lasted 73,000 years, until 123,000 YA. The following interglacial period continued until 110,000 YA, when the Earth again cooled. Between these times, the temperatures of the Earth were similar to those of today. When humans began migrating from Africa 60,000–50,000 YA the Earth's temperatures were again rising. Temperatures continued to increase until 30,000 YA when the Earth reversed course and descended into an extremely dry, cold period between 21,000 and 17,000 YA. At that time the sea level dropped 120 m below its present level. In the North, ice sheets covered Canada, Greenland, Northern Europe and parts of northern Asia. Approximately 17,000 YA, the Earth began to warm rapidly, the ice sheets retreated and the forests returned. These cyclic changes were due, as in earlier ice ages, to variations of the Sun's radiation on the Earth's surface plus the Earth's orientation relative to the Sun within the over lapping Milankovitch cycles.

Between 12,900 and 11,700 YA, the Earth suddenly descended into a short cooling period and another ice age. This event, however, was not due to the Milankovitch cycles but to the melting northern ice sheets which poured cold, fresh water into the northern Atlantic Ocean. The cold water blocked the flow of warm, saline water from the south. This weakened the down welling that powered the oceanic conveyor belt transporting warm waters of the south to the north. As a result, the northern hemisphere cooled. This interruption in the Earth's warming is referred to as the Younger Dryas interval.

For a number of years, it was thought that this interval was triggered by the impact of a large (1.5 km diameter) asteroid that struck the northwestern shore of Greenland 13,000 YA forming the Hiawatha Crater. Aside from destabilizing the ice and unleashing copious amounts of melt water, the impact would have also thrown dust and other light-blocking debris into the atmosphere cooling the Earth's surface. Recent data, however, indicates the impact is much older (58 MYA), arguing the trigger was perhaps a meteorite airburst instead of a direct strike or something else.

Holocene Epoch (11,700–0 YA)

The last 11,000 years of the Holocene Epoch have been the longest stable, warm period in the current interglacial span of the last half million years. The Holocene also corresponds to a time when humans expanded into nearly every region of Earth's surface and began to modify the environment. It includes all of human history including the transitions in human living from hunter gather, farming and industrialization to urban living.

At the beginning of the Holocene Epoch, approximately 4 million people had existed on the Earth. By 2021, there were 7.9 billion humans, which are increasing at the rate of 1.1% per year. This expansion in human numbers, from millions to billions, has significantly changed the planet. During the latter 200 years humans have affected the temperature of the Earth's atmosphere, land, water and environment. Carbon dioxide emissions, from the burning of fossil fuels, beginning with the industrial revolution, have contributed to Earth's warming, ocean acidification and habitat destruction. Scientists associated the increase in the average global surface temperature of 1 °C, measured between 1960 and 2020, to human activity, rather than due to any natural process. The appearance of human-made chemicals, minerals and radioactivity are now detectable throughout the sediments of the world's oceans, lake bottoms and ice fields. The current scientific community is thus considering the creation of a new epoch to reflect these human-initiated changes in the Earth's environment; suitably labeled the Anthropocene Epoch from Greek for human (anthropo) and new from (cene).

Part II
Humans

Chapter 4
Human Evolution

Apes to Homo Sapiens (8–0.2 MYA)

East Africa, Chimpanzees and Hominins

Approximately 30 MYA, during the Oligocene Epoch, tectonic plates began lifting and splitting an area of East Africa, forming a rift that extends north to south from Ethiopia to Mozambique. From 5.5 to 3.7 MYA, during the late Miocene and early Pliocene Epochs, these internal forces transformed what was once a relatively flat region into jagged mountains with deep valleys and plateaus. As the topography changed, so did the flora. Plant life ranged from that of a cloud forest, to a savannah and a desert. This region became East Africa's Great Rift Valley.

Closely related, during the interval 8–5 MYA, in the tropical forests of the Great Rift Valley, an extinct ape gave rise to both hominins and chimpanzees. At that time, the evolutionary difference between the hominins and the chimpanzees resulted from chimpanzees remaining more westwardly in the rain forests, as the region of East Africa was drying out; the hominins in the east, in contrast, adapted to the drying environment by evolving.

Because today's chimpanzees have changed little relative to the major changes in the hominins, their ape (and our) ancestor quite likely had characteristics similar to today's chimpanzees. These behavioral characteristics include competition for territory, food and sex. Their emotions range from, fear, jealousy or revenge to compassion, sharing, helping and altruism. They possess a keen awareness of social status and are constantly evaluating who is favored by whom. Humans and chimpanzees have 98.5% of their genes in common. As such, it is not surprising that these behavioral characteristics are also associated with the humans of today.

Many changes, however, did occur between the two species. One of the most critical was the placement of the larynx. In chimpanzees, the larynx remained high in the throat limiting the number of possible vocalizations. In humans, the larynx was lowered and created an air chamber to resonate sound, which ultimately allowed for spoken language.

Hominin Species

The first fossil evidence of a hominin was the Ardipithecus ramidus, who lived 4.4 MYA in Ethiopia, during the Pliocene Epoch. Ardipithecus was similar in size to modern chimpanzee and possessed a similar-sized brain (300–450 cc). For comparison, today's human brain size is 1200–1500 cc. Ardipithecus lived in trees and developed an early ability to walk upright on two feet. Over time, walking became a useful adaptation as the forests diminished and grasslands dominated, due to the continuous tectonic activity. Ultimately, this long-term drying motivated the evolution of tree-dwelling hominins, like Ardipithecus, to adapt to walking in the evolving grasslands.

Around 4 MYA, the genus Australopithecus afarensis emerged with well-developed bipedal walking and running ability. Lucy, also found in Ethiopia, was a famous example of this genus. She had a chimp-sized brain and was capable of walking long distances. Bipedalism served another function; it helped cool the body in the hot sun and allowed vision above the tall savanna grasses.

Australopithecus primary ate nuts, fruits, leaves and the meat of small animals. To avoid predators during the night, Australopithecus generally slept in trees, and likely used them as protection from predators. Because of their limited intelligence, they never fashioned stone tools. Lacking weapons, they were defenseless against Africa's larger predators. These hominins became the transition, between apes and humans.

Approximately 2 MYA, before the hominins of the Australopithecus genus became extinct; the genus Homo evolved from them. From 2.5 to 1.5 MYA, perhaps as many as 15 species, referred to as early Homo, were living in East Africa. With a brain capacity of 500–700 cc, only slightly larger than that of Australopithecus, their ability to make stone tools was primitive. They cracked stones together and used the resulting fragments as tools. Their fossil remains were first excavated in the Olduvai Gorge in the Rift Valley, along with lithics referred to as Oldowan.

2 MYA, Homo erectus also evolved from early Homo. Possessing a body and brain significantly larger in size, Homo erectus was anatomically similar to modern humans. Its brain size, however, was still only 70% of modern humans. Yet, with the greater brain capacity, they developed consistent patterns of mutual assistance and cooperation. Homo erectus became the first hunter-gatherer and the first to control fire. Fire not only provided warmth but also cooked food, which predigested the meal and made absorption of nutrients more efficient. The camaraderie around the fire likely contributed to the development of language and to new ideas for toolmaking.

Homo erectus refined the Oldowan tools into those known as Acheulean tools. These rock implements were made by carefully knapping increasingly thinner rock flakes that were then refined by the same process into more functional implements such as a pear-shaped hand axe. These tools represented the dominant stone technology for nearly the next million years. Ultimately, Mousterian tools created by later hominids replaced Acheulean tools. Again, two rocks were struck in such a way as to generate a large flake that was carefully knapped to form a thin, pointed rock now used as a knife, arrow head or spear point. These stone implements dominated through the Ice Ages. With these articles attached to shafts of wood to form spears, hominids and their descendants became effective hunters.

By 1.8 MYA, Homo erectus left Africa in multiple independent waves of migration and for the next 2 million years spread throughout Eurasia. Within the last million years, different hominin species evolved from Homo erectus. By 800,000 YA, Homo erectus gave rise to Homo heidelbergenis. Approximately 250,000 YA, Homo neanderthalenis (the Neanderthals) arose and populated Europe. In Asia, Homo denisovans dominated. Between 300,000 and 200,000 YA, the population of Homo erectus that remained in East Africa evolved into Homo sapiens. Approximately 70,000–60,000 YA, the Homo sapiens, now called humans, migrated from Africa and evolved to all the variations of humans populating the world today. By 200,000 YA heidelbergenisans, 50,000 YA denisovans and 28,000 YA neanderthals disappeared from the fossil record.

Need to Adapt

Numerous forces were at work to encourage larger brains that increased in intelligence, during this relatively short time period. During the middle of the Eocene Epoch, the world cooled and became drier. Between this trend and the tectonic uplift of the Rift Valley, the forests disappeared in East Africa. At the start of the Pleistocene glaciation, the alternating glacial and interglacial phases, were driven by the cyclical Milankovitch cycles. Within each of these phases, the climate oscillated between very dry and wet conditions; the global climate oscillated into an unstable state.

The Rift Valley was characterized by many large basins surrounded by high mountain ridges. If enough rain fell, streams from the highlands filled the basins, forming deep lakes. Generally, these lakes and riparian areas provided hominids with sources of water and food throughout the dry sessions. Many of these lakes, however, were not reliable as they filled or evaporated with climate variations. Small changes in the local climate, driven by cyclical, ice age conditions meant rapid swings in the water level of the lakes. This uncertainty created a strong evolutionary push on all life in the Valley.

Versatile hominins who adapted to the frequently changing climate conditions improved their chance of survival. Lewis Dartnell, in *Origins: How Earth's History Shaped Human History,* argues that it was these unique environmental fluctuations that drove the increased brain's size and therefore intelligence. New hominid species, generally with larger brains, appeared in the fossil records during the periods of changeable wet-dry conditions. Between 1.9 and 1.7 MYA, for example, five of the seven major lake basins within the Rift Valley continuously filled and emptied.

Concurrently, the number of hominid species, including Homo erectus, reached a peak during these fluctuations. These unstable conditions likely drove the evolving hominids from Africa in the first place.

Out of Africa Hypothesis

DNA studies as well as fossil evidence supports the "Out of Africa Hypothesis" for humans also. Modern humans, who inhabit the Earth today, evolved from just a single exodus out of Africa. It is estimated that such migrants numbered only a few thousand. In this hypothesis, modern humans first walked toward the Middle East and Arabian Peninsula around 70,000–60,000 YA, during a regional climate shift to wetter conditions. Upon reaching these new regions, their populations increased and their offspring continued to explore further new territory.

The Middle East provided a transition region prior to humans moving east, south and north. Migrants eventually entered Europe about 45,000 YA. Others traveled from Eurasia to India 50,000–45,000 YA. India, Southeast Asia and China opened to these walkers shortly after 50,000–45,000 YA. Continuing to explore, humans reached New Guinea and Australia 40,000 YA. The ice age conditions, with global sea levels more than 100 m lower than today's level, helped facilitate passage to the islands of New Guinea, Australia and Tasmania which were connected as a single land mass.

Those who traveled northward circumvented the Himalayas and eventually reached the northeastern tip of Asia 25,000 YA. As in Australia, the lowered sea levels resulted in a land bridge connecting Siberia to Alaska that provided a corridor a 1000 km wide. This Bering land-bridge, Beringia, was a barren landscape of an Arctic wasteland. There was barely enough vegetation to support woolly mammoths and steppe bison and the carnivores, like the saber toothed tigers that preyed on them.

By 20,000 YA, humans traversed this land bridge; and perhaps even earlier, as resent discoveries of human footprints between 23,000 and 21,000 YA in White Sands National Park, New Mexico suggest. They avoided the ice sheets found in North America, spreading into the land as far south as the Isthmus of Panama by 12,500 YA. Within another 1500 years, humans touched the southern tip of South America. With the exception of Antarctica, 11,000 YA humans had succeeded in occupying the entire world.

It is ironic that this last phase of human global expansion occurred during the harsh conditions of the last ice age. Without the lowered sea level, however, the land corridors connecting the continents would not have been available. During earlier ice ages, other animals had already crossed from the Americas to Eurasia. Importantly, for the future of human civilizations both the horse and the camel were among those animals that did so. Later, they became extinct in their place of origin, the Americas.

When Homo sapiens initially wandered into Eurasia, they encountered other species of hominids that had left Africa earlier. During the interactions with the Neanderthals in the Middle East, some degree of interbreeding took place and Homo sapiens picked up a component of their DNA, which was then carried throughout the world. Today, a small percentage of the Neanderthals' genetic code is embedded in all non-Africans. Similarly, cross-species mating occurred with another extinct hominid, the Denisovans in central Asia. The Denisovans were likely a sister species of the Neanderthals. By 40,000–24,000 YA, both Neanderthals and the little known Denisovans vanished from the fossil record, leaving only a genetic footprint behind. In the case of the Neanderthals, this occurred nearly as soon as Homo sapiens arrived in Europe.

Hunter-Gatherer (200,000–10,000 YA)

The increased brain size of Homo erectus did not automatically equate to broad mental capacity. Despite a larger brain and facility with fire, Homo erectus' essential skills never expanded beyond a limited repertoire of technologies. Their Acheulian stone axes, barely changed over a million years. Contrasting the limitations of Homo erectus and their dependents to those of humans highlights the differences. Unlike Homo erectus, humans continually adapted to their changing environments. Teamwork appears to have been the source of human social and technological creativity.

David Christian and coauthors suggest that the transformative event for humans, was language. Language enabled the efficient sharing of ideas, experiences and knowledge. In particular, the development of symbolic language where sounds are used for whole categories of ideas instead of simple identification of objects was key. Over generations, this learned information accumulated and is referred to as "Collective Learning". As a result, fire and tool technology combined with appropriate clothing, allowed humans to occupy every climatic zone from the warmth of the savannas and the tropics to the frozen tundra.

Initially, not much differentiated our ancestors from many of the other medium or large sized primate species. It is estimated that humans had a total population of no more than a few hundred thousand individuals, living in scattered communities, which grouped and hunted for food and resources together. Family or kinship relations provided the societal structure of these compact, Paleolithic, hunter-gather communities.

Such foraging communities needed an expansive territory to support themselves. This limitation restrained each group's population to typically 10–20 people and generally no larger than 150 members. There was little motivation to accumulate stores of food when what was needed could be attained from their environment. Moreover, when migrating, one carried only the most essential and portable goods. Accordingly, there was little differential in wealth among band members. Groups tended to be communal and egalitarian.

Although family bands lived separately for most of the year, they periodically rendezvoused where there was enough food to support temporary gatherings of hundreds of people. Once there, rituals, information, stories and mates were exchanged, which helped link these distant communities. Animistic beliefs prevailed and humans thought of themselves as a part of the natural environment.

Most humans still lived in Africa 100,000 YA. Climate change, conflicts with neighbors, over population, as well as environmental disruptions, motivated them to migrate. A steady flow of migrants explored and populated distant regions of the Earth.

Life was precarious and small groups could easily perish. Genetic evidence indicates, 70,000 YA the number of humans suddenly decreased to only tens of thousands, for example. This precipitous drop was attributed to the mega volcanic explosion of Mount Toba in Indonesia. Releasing enormous amounts of debris into the atmosphere effectively blocked sunlight globally for years. With less light, photosynthesis waned and Earth's flora declined. With less to eat, the number of herbivores also declined and both decreases impacted the subsistence of humans.

Eventually, human numbers increased and the cycle of communities was reestablished. By 30,000 YA, It was estimated that the human populations reached a half million; and by 10,000 YA it was up to six million. Although the technology of humans increased incrementally over these time spans, there was little evidence that energy was spent changing social dynamics or innovation; most of human energy involved population growth and the formation of new communities.

As humans expanded from Africa, their impact on the megafauna was profound. Unlike Africa where the animals had co-evolved with hominins; in Australia, Siberia and the Americas, many creatures were ignorant of how dangerous humans could be. In those regions, megafauna, like the 3 m tall marsupials of Australia, mammoths and mastodons of the Arctic and the herds of camels and horses in the Americas vanished soon after humans arrived. As recently as 10,000 YA, all mammoths disappeared.

Not only was the Earth's fauna affected by Paleolithic hunters but flora was as well. In Australia, New Zealand, parts of Eurasia and North America to create grasslands that would attract game to hunt, thickets and dense forests were deliberately burned. This practice was referred to as fire-stick farming. The scorched flora fertilized the land and thereby increased the number of usable plants and thereby animals. An unintended result of this practice was that the rare eucalyptus trees in Australia, which were particularly resistant to fire, spread significantly.

Chapter 5
Agricultural Revolution

Transition to Agriculture

Prior to 17,000 YA, foraging humans had survived two distinct ice ages. These fluctuating and harsh climatic conditions made it difficult for a stationary life style to be based on agriculture as a food source. The foraging strategy of the hunter-gatherer was a better approach to surviving a changing environment. The warming of the Earth at the end of the Pleistocene Epoch 11,700 YA and an increasingly stable climate, however, precipitated change. At this time, warmer temperatures meant more oceanic evaporation, which in turn, increased rainfall, allowing forests and grasses to flourish. Temperatures became similar to those of today.

By this time, 5–8 million foragers, had organized into tribes. Each tribe developed its own culture and language. The Earth was becoming increasingly populated. Foragers covered so much of the Earth's surface that some regions felt overpopulated. Foragers became aware of sustainability and how many individuals the land could contain. Ten thousand years ago in Europe, for example, 10 km² of land could support one human. The changing environment and improved climate, combined with an increasing population contribute to the explanation of why foragers transitioned to agriculture, nearly simultaneously around the world. Farming could feed higher population densities. This drastic change in life style represented a fundamental transformation in human development.

Based on archeological evidence, the transition to agriculture first occurred in the Middle East in the region referred to as the Fertile Crescent. This broad swath of arable land extended from the hills of what is now southeastern Turkey, west to Egypt and south through Syria, the Euphrates and Tigris River valleys to the Persian Gulf. This productive region between those two rivers, in present day Iraq, is referred to as Mesopotamia. The improved climate that began after 17,000 YA (aside from the Younger Dryas interval) was ideal for growing wheat, barley and other grains (all of which are species of grasses). As foraging ancestors began eating more grains, wheat in particular, the practice of agriculture spread.

T. Sanford, *A Whirlwind History of the Universe and Mankind*,
https://doi.org/10.1007/978-981-97-2674-5_5

The foraging humans gradually modified their nomadic existence to settle in seasonal or, in some cases, permanent locations. This was the case for Natufians who lived in the Levant (the coastal region of Syria, Lebanon and Israel) 14,500–11,500 YA. They were among the first to adopt a semi-sedentary way of life. They built small, permanent villages of round stone houses, and used the abundant resources of the land for sustenance. They were referred to as affluent foragers. Later, the Natufians cultivated and processed native grasses into edible cereals. Gathering was not completely abandoned. Nuts and fruits still augmented their diet. Gazelles and asses were hunted or kept in pens to provide protein.

The dawn of agriculture had far reaching ramifications. With agriculture came related tools. Scythes were invented with blades of sharpened flint for harvesting the wild wheat. Stone mortars and pestles were created to grind the cereals. Stone granaries were constructed to store excess grain. The descendants of the Natufians discovered that productivity increased when seeds were sown rather than scattered. Tools were used to dig, hoe and plough the ground. Eventually, fields were weeded, irrigated and fertilized. As cereal cultivation increased, the time to gather wild species of fauna and hunting decreased or wasn't necessary. These innovative offspring were officially the first farmers.

Another ramification of farming was human reproduction. The output of food from farming fed more people than foraging. Populations grew. Extra children were desirable as they could contribute to food production. Nevertheless, farming was difficult and foraging survived alongside farmers for thousands of years. In regions more suitable for crops, the advantages of farming meant that eventually farming superseded foraging. Typically, an area of fertile land could, even with primitive farming techniques and tools, produce ten times more food than foraging. Once farming became the main means of food generation, foraging skills were mostly forgotten. With larger populations to feed, foraging was no longer sustainable or practical.

Between 11,000 (9000 BCE) and 5000 (3000 BCE) YA, in addition to the Fertile Crescent, farming villages evolved independently throughout the world. Other river areas, such as the Nile in Egypt, Yellow and Yangtze in China, Indus of Pakistan and Ganges of India, all attracted farmers. During these 6000 years, most farming communities, bonded by traditional rules of kinship and remained independent with minimal interaction among neighboring villages. Initially, these early farming settlements were relatively equalitarian.

Early Villages

Over time, some villages (composed of hundreds of people) eventually grew to become towns (of thousands of people) or city-states (of tens of thousands). Reasons for that growth varied. Some had reliable sources of water, while others were on trade or migratory routes. Still others occupied strategic locations or became important ritual centers for surrounding villages. Three early examples follow.

Jericho

Jericho was the oldest and almost continuously inhabited human community known to mankind. Located 263 m below sea level in the fertile, alluvial Jordan River Valley, a reliable spring still supplies water to residents to this day. Affluent Natufian foragers occupied Jericho 14,000 (12,000 BCE) YA. Later, forager-farmers constructed round houses of mud-brick and grew wheat and barley and domesticated sheep. The mostly identical houses had their own grain storage areas and were clustered by kinship groups.

Between 10,350 (8350 BCE) and 9350 (7350 BCE) YA, the village transitioned into a town of approximately 3000 farmers spread over an area of 10 acres. The town was protected by a 4-m stone wall with an enormous round stone tower and massive ditch. The tower stood 8.4 m high, was 9 m in diameter, with 1.5 m thick walls and a staircase led to the roof. The wall and tower were one of the first known massive construction projects that required cooperative labor and the need for protection. Around 3580 (1580 BCE) YA Jericho was destroyed, which may be related to the Biblical story of Joshua. However, archeological evidence is inconclusive.

Catalhoyuk

Catalhoyuk was an early agrarian settlement, that developed 9300 YA (7300 BCE) in the Konya plain of southern Turkey. Today, this once fertile plain is now a dry region. But then, with its multiple flowing streams, formed reed marshes. Its alluvial soil and water made the area ideal for farming. The town itself was believed to be a consolidation of a number of earlier nearby villages; by 8200 YA (6200 BCE), it contained up to 8000 people. The multistory structures were similar to the famous Taos pueblo in northern New Mexico. Each house was made of timber framing and mud bricks with rectangular rooms adjoining its neighbors on all sides. Like Taos Pueblo, room access was through the roof via ladders. The inhabitants cultivated cereal and vegetables and raised herds of sheep and goats that provided meat and milk.

The people of Catalhoyuk also traded obsidian, found 128 km away in the volcanic region of Cappadocia. Obsidian was used to flake hard, sharp, stone projectile points for hunting and tools for farming. In many of these early agrarian towns there was no significant evidence for emergence of leadership or the religious elite.

Gobekli Tepe

Gobekli Tepe was yet another example of an early settlement. It was located at the highest point of a mountain ridge, 15 km northeast of present-day Urfa in southeastern

Turkey. Gobekli Tepe was first occupied during the time of Natufian villages and then occasionally between 12,000 (10,000 BCE) and 9000 (7000 BCE) YA. It was contemporary with Catalhoyuk, but very different.

Gobekli Tepe represented the technical and organizational ability of early affluent foragers. It consisted of twenty stone circles with approximately 200 carved stone pillars imbedded in or within the walls. Given that some of the pillars were well over 5 m tall and weighed nearly 20 tons, this town likely served as a cultural or religious hub. This center was thought to be connected to domestication of wheat and served as an observatory. To be built, it required thousands of foragers belonging to different tribes, working together, over an extended period of time. Gobekli Tepe expressed a new level of social complexity.

Continued Development

While the social, economic and political structures necessary to develop cities would take thousands of years, agricultural innovations continued unabated. Once agriculture based communities proliferated, there was an increase in the control of animals. By 8000 BCE, sheep were domesticated in the Taurus Mountains of Turkey and goats in the Zagros Mountains of Iran. At the same time, cattle were trained from wild aurochs in the Near East and India. The common pig, derived from wild boars, was found in Europe and Asia by 8000–7000 BCE. Chickens appeared in south Asia around 6000 BCE. Horses were present in the Ukraine and donkeys in Egypt around 4000 BCE.

Initially farmers ate their domesticated animals as soon as they matured. The meat, hide and bones were all used for food, shelter and tools. Wheels and plows emerged by 5000 BCE in Afro-Eurasia and humans quickly realized they could hitch their animals to carts and plows for more muscle power. Gradually, by 4000 BCE, farmers perceived that livestock could also provide long-term needs beyond carrying loads and pulling plows. Goats and sheep, and later cows, yielded and allocated calcium rich milk, cheese, butter and yogurt, which remained stables over extended periods of time.

Not all regions were fertile and bountiful. Where there was minimal arable land, humans raised herds of livestock that roamed over great distances. Cattle, goats and sheep thrived on sparsely grassed land that otherwise was unsuitable for farming. Mounted on horses, men managed large herds grazing over the vast grasslands of central Eurasia and opened the area for habitation. Nomadic pastoralists across the Eurasian steppes played an important role in promoting trading routes among the nascent city states on the fringes of the steppes.

By 6500 BCE, the art of pottery was perfected in Mesopotamia, which solved the problems of transporting and storing grains and liquids and ceramics greatly enhanced culinary skills.

By 5500 BCE, soft metals like copper, silver and gold were being worked. Bronze (a mix of 10 parts copper to one part tin) appeared 1500 years later in 4000 BCE.

By 3500 BCE, the first signs of its usage materialized in Mesopotamia. The years between 3500 and 1000 BCE represented the Bronze Age.

From these many agricultural and metal working developments that spanned thousands of years, city-states began developing by 3500 BCE, again originating first in Mesopotamia. These urban centers became the next major advancement in communities.

Early City-States in Sumer

By 6000 BCE, farmers were migrating south from the hills of southeastern Turkey to the fertile plains of Mesopotamia, bringing their cultivated cereal crops. The region was warm and wet and the area of lower Mesopotamia with its deep alluvial soil was extremely fertile. Population growth and the drying of the hill region spurred others to move southward as well. The farming was ideal and the population grew, forming many villages along and between the Tigris and Euphrates Rivers. The desert restricted these river populations from expanding into the country side.

By 3800 BCE, the climate cooled, the rains became less reliable and the land began to dry. In response, the villages along the two rivers combined their resources to form larger communities, from which they constructed greater and more effective irrigation systems. The increased population, and the building and maintenance required of the irrigation canals, subsequently promoted a more centralized administration.

Communities, with a stratifying social structure supported by farms, coalesced into cities or city-states.

The city-state comprised the geographical area of a central city and its surrounding territory. Together they formed a single, self-governing political unit, with its capital as the focal point of the state, thus, becoming the first urbanized centers of commerce.

By 3500 BCE, numerous city-states emerged. The major purpose of these city-states was to construct and maintain their life blood—the irrigation canals used both for agriculture and transportation. By 3000 BCE, more than a dozen city-states including Uruk, Ur, Eridu, Kish, Nineveh and Nippur existed. The collection of these cities along the lower reaches of the Euphrates and Tigris Rivers was referred to as Sumer.

Uruk, which was a port on the Euphrates bordering the swamp of the river's delta, was often referred to as the first city in history. In 3500 BCE, Uruk already had a population of 10,000 people. Two hundred years later in 3300 BCE, it swelled to 50,000 and covered an area of 2.5 km^2, reaching its zenith in 2900 BCE. At that time, the city was crossed with numerous irrigation and transport canals surrounded by 10 km of burnt-brick walls, protecting perhaps the largest human settlement on the Earth.

Uruk was composed of three general areas. One walled section included temples, palaces (after 3000 BCE) and the residences of the citizens. A second section was a commercial area housing shops of foreign vendors. The third section was an outer ring of farms, gardens and livestock fields. Within the city walls, the people engaged in hundreds of different occupations, including: priests, officials, scribes, potters, cooks, bakers and metal-smiths.

For 500 years, the people of Uruk led the development of Sumer. Uruk's boats traded up and down the river systems of Mesopotamia and as far away as India. By 3000 BCE, Uruk had dissolved into just another city of Sumer. Other cities, ruled by kings, began controlling their own territory and waterways.

Each city was generally characterized by the construction of a monumental ziggurat. These stepped, mud-brick and tiered pyramids held temples that honored various Gods to protect the city and bring prosperity. The initial purpose of the ceremonial centers was for storage of grain, places for decision making, rituals and processions. Priests were thought to be in charge of building them, as well as storing the harvested grain and distributing it.

In Ur, the top exterior space supported a temple and allowed the priests to monitor the sky. The Ur ziggurat became associated with the sky God, the father of all Gods. Within Ur, next to the ziggurat center, was a section connected to the sky God's daughter, Inanna, the Queen of Heaven, goddess of love making and later referred to as Ishtar. The subsequent recording of such observations in Uruk led to the emergence of Uruk as a center for astronomy.

The Mesopotamians believed the world was created out of the merging of fresh water with the salty ocean, one male and the other female. The forces of nature became Gods who behaved as humans, except that they were immortal. Storms, wind, fire, rivers, and mountains were regarded as living beings, filled with spirits. They were pessimistic about life after death, where one existed in a place of perpetual darkness. The Gods eventually became treated as the upper rungs of their society. Social cohesion was promoted through religion, which helped legitimize a ruler's authority.

The advent of cities necessitated social constructs beyond the simple division of labor. A professional class of bureaucrats became necessary to keep the system running. As labor specializations increased, so did social inequality. Most people were free commoners who farmed their own land or were dependents that owned no land but worked for others. Some were slaves, often prisoners of war, who worked as domestics in wealthy homes and as labor in the fields. As early as 4000 BCE warfare was present in Sumer with water and land typically being fought over.

The earliest writing was found in a large Inanna complex of Uruk dating to 3500 BCE. It was initially motivated by the need to keep track of inventories of food, animals, goods and trade items. Writing was made using reed sticks imprinting wedged-shape incisions on wet clay tablets that hardened, using Uruk's natural resource of mud and reeds and was referred to as cuneiform. Cuneiform continued to be used until 700–600 BCE, well past the decline of Sumerian influence, at which time Aramaic (a region of Syria) replaced it.

Initially, scribes incised stylized pictures of an object. Abstract numerals appeared alongside the pictographs by 3100 BCE. The number system used was partly based on the decimal system and partly based on the numeral 60. The evolution of true writing with a combination of symbols representing objects, ideas, syllables or sounds developed from 3300 to 2700 BCE. The surviving texts were written in 2100 BCE and speak of King Gilgamesh, who ruled Uruk around 2750 BCE. The Sumerian oral myths were written in *The Epic of Gilgamesh: Gilgamesh's accomplishments and his adventures*. The Epic was a timeless tale of friendship, the relationship between men and Gods, the meaning of life, the fear of death and the impossibility of immortality. The year 2100 BCE designates humankind's first written history.

The Sumerians excelled in their understanding of astronomy and mathematics. They developed a calendar based on 12 months in a year, a day that was divided into 24 h, an hour that was divided into 60 min and a minute that was divided into 60 s. In mathematics, a circle was partitioned into 360°.

In 2250 BCE, an unprecedented dry spell with reduced rainfall altered the once favorable climate of Sumer. After millennia of use, the alluvial soil became less fertile as years of extensive irrigation and evaporation left a salty residue. Between of the lack of rain and the increasing soil salinity, by 1700 BCE, crop yields decreased to 66%. Sumer became an impoverished region that descended into insignificance as other civilizations rose elsewhere. Nevertheless, Sumer, as being the first collection of city-states, provided a template for future city-states. It represented the transition from agrarian villages to an agrarian civilization.

Early Empires

Increasing populations and the competition for resources, meant conflict was inevitable. The need for more water or land, power and prestige among leaders provided the incentives for expansion and control over neighboring states.

In 2560 BCE, the King of Ur overthrew the King of Uruk. By this time professional armies were employed. A few hundred years later, Sargon of Akkad who ruled from 2334 to 2279 BCE, conquered all of Sumer from the Persian Gulf to the northwest end of the Euphrates River. Sargon boasted that he controlled the entire world. Referred to as the Akkadian Empire, it collapsed by 2150 BCE due, in part, to the drying climate. His expansion formed one of the first empires.

In 1810 BCE, Hammurabi succeeded in again uniting Mesopotamia, making Babylon along the Euphrates River the capitol of his empire. Hammurabi is remembered for introducing the first written law code and the expression "An eye for an eye."

Concurrent with the forming Mesopotamia Empires, to the north a proto-Hittite group migrated from the Balkans to Anatolia around 2750 BCE. Eventually, that group formed the Hittite Empire (1800–1200 BCE) centered at Hattusa in north central Anatolia. The Hittite Empire reached its zenith by mid 1300 BCE, when it expanded throughout Anatolia, the northern Levant and the upper part of Mesopotamia. The Hittites used a new weapon of the age, the horse-drawn chariot. Circa 1375 BCE, the Hittites captured Babylon before retreating to Anatolia.

Once Hittite power declined, Assyrians (originally centered at Assur in the upper part of Mesopotamia) occupied the fluctuating power gap. Like the Hittites, the Assyrians effectively used horse-drawn chariots. Although the Hittites developed techniques for working iron, the Assyrians were the first to arm themselves with iron weapons. The Assyrians formed the largest agrarian civilization in early history. At its peak, the Neo-Assyrian Empire (911–609 BCE) spread from Egypt to Cyprus, to Iran in the east, Armenia and Azerbaijan in the north and the Arabian Peninsula in the south.

During this period, Akkadian and Aramaic became the official languages. The Neo-Assyrian capital, Nineveh (today Mosul, Iraq, on the Tigris River), was considered the largest city at the time. After an unending series of civil wars, it too began to falter. In 609 BCE, Assyria was crushed by the Babylonians and Medes (from the Zagros Mountains), after which it never recovered as an independent state.

The shattered remains of the Assyrians were replaced by the militarized Persians of Iran. In 560 BCE, Cyrus the Great, founded the Achaemenid Empire. Under subsequent rulers this empire expanded. By the reign of Darius I, who ruled 521 to 486 BCE, the empire stretched from the northern Libyan shore, through all of Egypt to Asia's Indus Valley. The Achaemenid Empire exceeded the territories of the Assyrians and then became the largest empire.

To manage this vast area, the Persians introduced a new form of governance. Maintaining the multicultural empire required a delicate balance between local and centralized rule. At the local level, no attempt was made to institute imperial laws (laws issued by the empire). Governors presided over semiautonomous provinces. To promote coherence, a network of roads were constructed with the "Royal Road" extending over 2560 km from Sardis near Ephesus on the Turkish shore, to Susa at the foot of the Zagros Mountains near the head of the Persian Gulf.

In general, empires condensed many small cultures into a few more sizable ones. Within the larger parameters, the population, goods, technology and ideas spread more rapidly. Although formed in bloodshed, empires standardized writing, oral language, laws, weights, measures and money. Subsequently, empires improved living standards and made governments more efficient.

Other Early City-States

At the time agrarian villages were forming in the Fertile Crescent, other regions along a similar latitude, were also developing city-states in Africa and Asia. The drying climate simultaneously pushed some of these early, scattered agrarian communities into nearby river valleys, where a more complex social organization took place. These included, Egypt's Nile River Valley to the west, Pakistan and India's Indus River Valley to the east and China's Yellow and Yangtze River Valleys further east.

Nile River Valley

The North African climate around 9000 BCE, unlike today, was temperate. The region, with its adequate rainfall and comfortable temperatures, grew enough wild grains to support humans and wild cattle. Over the next millennia, humans cultivated these plants along with cotton and watermelons and domesticated animals. Goats and sheep arrived from the Fertile Crescent to provide milk and protein. Villages run by hereditary chief-kings ruled the grassland regions by 5000 BCE.

As happened in the hills of southeastern Turkey, the climate began to dry, pushing the population into the Nile River Valley. Initially, the fertile soil of the Nile needed no irrigation, but as the population increased and spread beyond the flood plain, irrigation became necessary. By 4000 BCE, the Nile was lined with agricultural villages from Nubia to its delta on the Mediterranean Sea. As in Sumer, irrigation systems required cooperation and leadership to function and allowed trade up and down the Nile.

Five hundred years later, rainfall declined and villages combined into chiefdoms. Competition between chiefdoms increased. Circa 3500 BCE, Egypt merged into two kingdoms: Lower Egypt, the delta region of the Nile and Upper Egypt, the territory south of the delta to the first river cataract. Menes, the king of Lower Egypt, ultimately united both kingdoms in 3100 BCE. To facilitate unification he founded the first capital of Egypt, Memphis, at the spot where the two kingdoms abutted each other not far from today's Cairo.

Memphis was built on an island in the middle of the Nile and was easily defended. Navigation north and south along the Nile was the link that bound the two kingdoms together, with the north-flowing current that flowed against the reliable south-blowing winds. Unlike Mesopotamia, Egypt used the Nile's summer floods to regenerate the soil annually. The floods removed any accumulated salts from the frequent irrigation.

Menes, considered to be the first true pharaoh, initiated a series of 30 royal dynasties. The dynasties were families who ruled until there were no heirs or until there was a successful revolt. There was a tradition to consider kings to be divine that dates back to the villages in North Africa. Now they were called pharaohs. Isolated by surrounding deserts and self-supporting, thanks to the Nile, the pharaohs ruled Egypt for over three millennia, including intermittent foreign invasions, as a region of enduring pharaonic religion.

Over this extended period of time, Egypt matured through three major periods that are referred to as the Old Kingdom (2700–2200 BCE), the Middle Kingdom (2050–1800 BCE) and the New Kingdom (1550–1100 BCE). The population prior to this dynastic period was estimated to be under a million people but grew to about three million by the New Kingdom. These three kingdoms were followed by the Late New Kingdom that lasted until 343 BCE when Alexander the Great conquered Egypt in 332 BCE. The Old Kingdom is often referred to as the Age of the Pyramids. The three famous pyramids of Giza were built as tombs for the pharaohs during that time.

Between the kingdoms control collapsed. The end of the Old Kingdom ended with a drought, low-level floods, poor harvests and famine. These factors resulted in anarchy and civil war. The Middle Kingdom formed when the rulers of Thebes (Luxor) defeated the pharaohs of the north and made Thebes Egypt's Capital. The Middle Kingdom ended when the Hyksos, a Semitic Asiatic people, invaded the Delta region and defeated the Egyptian forces. At this time, the Egyptians fought on foot, using stone and copper weapons. In contrast, the Hyksos had the advantage of bronze weapons and horse-drawn chariots. Eventually, the Thebans reunited the country and the Hyksos were expelled to form the New Kingdom. This period is referred to as the Golden Age of Pharaohs. Most of the monumental structures at Thebes, the religious and political center of the New Kingdom, date from this period. The old capital of Memphis was left to attend to administrative and other day-to-day issues.

The Great Pyramid of Khufu in Giza (near Memphis), captures the Earth's attention to this day. It is both the oldest and the largest pyramid and has the distinction of being the only one of the seven wonders of the ancient world still standing. Completed in 2570 BCE, at a height of 146 m, it was the tallest human-made structure on Earth until the Lincoln Cathedral, which was built in 1311 CE in England. The Great Pyramid was constructed using 2.3 million limestone blocks, each with an average weight of 2.5 tons. The blocks were precisely cut to within a mm; the mean opening between the joints being only 0.5 mm. The pyramid was aligned north-to-south with an accuracy of 0.05°. This monument illustrates the Egyptian's command of mathematical and engineering skills, as well as, the social and political organization needed to build such a complex structure.

More than 500 years later, south at Thebes, other temples would highlight the advanced social and technical skills of the Egyptians. There, structures at Karnak and Luxor, during the Middle Kingdom, were examples of Egyptian engineering, which existed through generations of pharaohs. Karnak covered an area of 6 km^2 and was the largest temple complex in the world, until the Hindu Temples of Angkor Wat were built in 1150 CE, more than 3000 years later.

Karnak's Great Hypostyle Hall was most impressive, with its 134 large, papyrus-shaped sandstone pillars arranged in rows of 16 columns. The center 12 columns stood 21 m high, supporting a ceiling of blocks that created the largest enclosed area until the arch was invented. The temple's 20 obelisk tips, coated with electrum (an alloy of gold and silver) reflected sunlight throughout the complex. The tallest obelisk was a 30 m high slab of granite weighing 450 tons and was erected under the reign of Queen Hatshepsut.

Karnak's granite was excavated from the banks of the Nile near Aswan and floated down the river on boats to the construction site. The smaller temple of Luxor, with many of the same features as Karnak, was connected to Karnak by a 5 km road lined with carved stone sphinxes acting as sentinels of the causeway. The exterior and interior of both complexes were inscribed in hieroglyphs, a stylized form of Egyptian writing.

As in Mesopotamia, the initial need for writing was for accounting purposes. Writing began as simple pictographs of objects and was then transformed into symbols representing ideas and sounds. This form of Egyptian writing called hieroglyphics, was widely used and decorated tombs and temples by 3100 BCE. Unlike

Sumerian incising in wet clay, by 3000 BCE, Egyptian papyrus was employed. It was created from a reed growing in the Nile's marshes. Between 2800 and 2600 BCE, a simplified form of hieroglyphics called hieratic appeared and later a vernacular script called demotic was used. Hieroglyphics continued to be used until 300–400 CE when it was replaced by Arabic. The hieroglyphics were indecipherable until the Rosetta Stone was discovered in 1799. This stone, with its incised message in Greek, demotic, and hieroglyphs, finally enabled scholars to decode the Egyptian writing.

Egyptians practiced a polytheistic system of beliefs that centered on respect for truth, balance, order, harmony, law, morality and justice. These traits were exemplified in the Goddess Ma'at. Pharaohs were believed to possess divine power and acted as an intermediary between the people and the gods for the well-being of all. Egyptians were obligated to sustain their gods through offerings, rituals and ceremonies. They sought order in the universe that assured the annual flooding of the Nile and a smooth transition after death.

Egyptians believed in life after death. Reaching a happy afterlife was not a simple matter. Acting fairly and honestly throughout one's life was key to a successful transition. This behavior was judged when the deceased's heart was weighed against the feather of truth by the God Anubis in the Hall of Ma'at. With a light heart one easily passed into heaven for eternity. But, if your heart was heavy due to poor behavior, the God Ammut would appear and eat you. Aside from the possibility of being eaten, this belief was in stark contrast to the Sumerian's more pessimistic view.

To assure a smooth transition to the next world the Egyptian society was preoccupied with a set of ritualized preparations that included mummification. The state dedicated large amounts of physical and monetary resources to facilitate this passage by the construction of temples and tombs and the mummification of important leaders. Mummification and the placement of one's body in an entombed stone sarcophagus was all an attempt to secure a successful journey into the next life.

During the Old Kingdom, construction of tombs within pyramids, such as at Giza, was the typical form of interment for the nobility. In contrast, during the later kingdoms, when power had shifted upstream to Thebes, sarcophagi were placed into crypts carved in the west facing cliffs of the Nile at Thebes. The crypts were sealed by rock to prevent theft of the funereal offerings and goods necessary for the afterlife. This was the case of the well-known King Tut, born in 1336 BCE, during the New Kingdom. His tomb and golden treasures had remained untouched until its discovery by the British archaeologist Howard Carter (1874–1939) in 1922.

Indus River Valley

Like the inhabitants of the Fertile Crescent, the art of cultivating also occurred in the Indus River Valley by 7000 BCE. Seeds were sown in the fall, after the flood waters receded, and harvested in the spring. By 5000 BCE, cotton was cultivated for cloth production. They also domesticated water buffalo, cattle, goats, sheep and chickens.

As in Mesopotamia and Egypt, the climate changed and the land became more arid after 4000 BCE. People migrated into larger communities and cities in the general region of the Indus River and its tributaries. There, the various cultures of the Indus Valley became collectively known as the Harappan Civilization.

Two Harappan cities on the Indus River have been extensively excavated: Harappa and Mohenjo-Dara. Harrapa was located near the foothills of the Himalayas while Mohenjo-Dara was founded further downstream. Both arose about 3200 BCE. At their maximum extent 2500–1900 BCE, Mohenjo-Dara had an estimated population of 35,000–40,000 people. Harappa had somewhat less. Each city's footprint was approximately 2.6 km^2.

The Harappan civilization began to decline in 1800 BCE when rainfall lessened and subsequent droughts became frequent. Additionally, it is theorized that migrating Yamnayas, an Indo-European tribe of nomadic cattle herders, also invaded from the steppes and conquered the Indus Valley.

Though separated by many kilometers, Harrapa and Mohenjo-Dara were surprisingly similar. Each was a walled city with streets designed in a grid pattern aligned with the cardinal directions. Both contained market places, small temples and public buildings. Each had a large granary with a site for collecting and redistributing taxes paid in grain. Many of the houses had wells, indoor bathing facilities and elaborate underground drainage systems with pipes to carry away the waste water under the streets. The houses were constructed from sun dried mud bricks, while baked bricks were reserved for bathrooms, wells and drains. Timbers supported flat roofs. Only Mohenjo-Dara had a great bath used for religious purposes and purification of bathers.

The Harappan civilization was noted for other advances beyond its architecture. The citizenry had mastered the art of metallurgy, producing items made from copper, bronze, lead and tin. They were a literate civilization whose written language has yet to be translated. The Harappans traded with their neighbors in all directions, including Mesopotamia, Persia, Afghanistan and central Asia to the north. By 2300 BCE, they were sailing down the Indus River, clockwise around the Iranian Plateau on the Arabian Sea and north through the Persian Gulf to the ports in Sumer.

The Harappans were grounded by religious beliefs that, uniquely, had no political overtones. Their beliefs were similar to modern-day Hinduism. The Indus people worshipped a Father and Mother God associated with creation and procreation. They also practiced yoga and meditation. Unlike other city-states, there was no archaeological indication of a political hierarchy, usually indicated by palaces or large temples. There was little depiction of kings, soldiers or warfare in its art, discovered during archaeological excavations.

There was speculation that a caste system, developed from 3000 BCE onward, would explain why the Harappan civilization spread without warfare. People would be encouraged to remain content with their present station in life, knowing that if they accepted it gracefully, they would advance to a better position in the next life.

Yellow and Yangtze River Valleys

Two great city-states occurred in present day China's Yellow River Valley in the north and the Yangtze River Valley in the south. Both appeared to develop simultaneously but initially, independently. More is known about the Yellow River cities than about those of the south where excavations have been limited. In the north, along the Yellow River, people made steady technological advances over a long period of time. Living on the plateau and central plain of the Yellow River, people cultivated millet by 7000 BCE. Agricultural villages clustered in 5000 BCE. By 2700 BCE, the technique of silk processing from silk worm cocoons was perfected and used to weave cloth. Metallurgical skills were imported from the west by 2500 BCE just as walled settlements arose. By 2000 BCE, wheat and barley were cultivated along with hemp, which was used for clothing.

As the climate changed in the west, China was also becoming cooler and dryer. The change motivated people to move into the fertile Yellow River Valley. As a result, from 3000 to 2000 BCE, the population of the farming communities expanded significantly. The Longshan culture in the middle and lower Yellow River (3000–1900 BCE), with its rammed earth walls surrounding each neolithic settlement, was an early turning point to the emergence of the dynastic states that followed.

Eventually, an urban, literate Bronze Age culture known as the Shang Dynasty (1600–1046 BCE) controlled the territory between the lower reaches of both the Yellow and Yangtze Rivers. Its population approached 14 million people. Over 500 years, the Shang Dynasty was ruled by a succession of 30 different emperors. During the reign of Wu Ding (1250–1192 BCE), writing first appeared as symbols for objects. Later, writing represented abstract and complex ideas; however, it never included phonetic or alphabetic components. Written language expanded to include over 4000 characters and has been used continuously ever since.

There were noticeable differences between the development of cultures in the Yangtze and Yellow River Valleys. In contrast to the north, in the subtropical south of the central Yangtze River Valley, wet-rice farming expanded by 6500 BCE. Wild rice had always been a part of foragers' diet but now it was a staple. By 4000 BCE, walled towns with moats appeared. By 2500 BCE, social differentiation evolved, as indicated by the varying sizes of houses. Over time, the Yangtze Valley culture generated a civilization that competed successfully with the Shang Dynasties.

Other Regions

Connecting the eastern to western civilizations were those developing in central Asia. There, a Bronze Age civilization known as the Oxus evolved between 2400 to 1600 BCE, whom the Shang traded with. In turn, the people of Oxus also traded with India, Sumer and the pastoralists of inner Asia whose economy was based on sheep and cattle. These connections laid the preliminaries for future Afro-Eurasian trading networks.

Early agrarian civilizations also developed in other regions of the world. These, however, occurred later in time than those of Afro-Eurasia. Two notable cultures were in the Americas: one in Mesoamerica and the other in South America. In Mesoamerica, agrarian communities existed by 2000 BCE. With no large domesticated animals, their economy was based solely on crops: corn, beans and squash. In the Andean Mountain region (currently Peru and Bolivia), wide-spread cultivation matured between 2500 and 2000 BCE, where potatoes became a staple of the diet. They, raised herds of alpacas used for wool and llamas for meat. In the coastal regions, farmers planted beans, sweet potatoes, peanuts and cotton, while fish and other sea life supplemented their diet.

Advantages of Old World Over New World

By the end of the last ice age, people had populated all corners of the Earth, except Antarctica. Settled, agricultural communities in the Old World (Afro-Eurasia) led to the early urban civilizations discussed. The remainder of this book focuses on developments in this region, where a more structured civilization first occurred. However, it is worth a comment on why civilization evolved 5000 years earlier in the Old World relative to those in the New World (the Americas).

To foster the initial transition to an agrarian society, the Old World had several advantages relative to the New World. Jared Diamond, the author of *Guns, Germs, and Steel: The Fates of Human Societies*, and other authors have presented an elegant explanation. One factor was the uneven distribution of wild grasses to produce cereal throughout the world. Of the 56 grass species suitable for nutritious domestication, 52 were found in the Old World but only 10 in the New World. Another major advantage was the orientation of the land masses. In the Old World, the land spreads west to east along almost the same range of latitudes and their associated climates. Plants in one region were therefore easily cultivated in other areas. Wheat cultivation, for example, spread from the southern hills of Turkey, south through Mesopotamia, north into Europe and east into India and China. In contrast, the North to South America, connected only by the narrow Isthmus of Panama, which spanned multiple latitudes making adapting plants from one region to another difficult, if not impossible.

The distribution of large animals suitable for domestication was also uneven. Eurasia contained 72 of the 148 large, wild animal species, whereas only 24 roamed the Americas. Not all wild animals were suitable for domestication. The characteristics that made domestication possible were the ability to breed in captivity, a docile nature and availability of food. Of the 72 Eurasian animals that met these criteria, only 13 were able to be domesticated. In the Americas, of the 24 animals, only two could be domesticated, the llama and alpaca of South America. In contrast, the goat, sheep, pig, cow, horse, donkey and camel were among the many domesticable wild animals found only in Eurasia. They were crucial for the development of agriculture and transportation, as well as for their secondary food and clothing products. This gave those living in the Old World a huge boost when developing their civilizations.

Ironically, the horse and camel evolved first in the Americas. Both migrated into Asia, during the last ice age when the Bering Strait was bridged. Important for transportation, they apparently were hunted to extinction by the Paleo-Indians who were crossing the bridge in the opposite direction into the Americas.

In addition, the development of language and the wheel advanced the civilization in the Old World relative to the New World. The writing systems of Mesoamerica, a combination of pictographs and hieroglyphics, never moved into North or South America or evolved into alphabetic form. This non-development limited the dispersion of knowledge. Likewise, the wheel, invented in Mesoamerica as parts of toys, was never conceptualized as a tool to be used for transportation nor was the idea exported to the north or south.

Chapter 6
Mediterranean Development

The path from the creation of the universe, through to the early development of civilizations that eventually led to today's conception of a world, historically passed to the Mediterranean region. Before considering the many city-states that developed there, a brief description of how the Mediterranean Sea and its surrounding geography formed is appropriate. This discussion connects Earth's tectonic progression to the topography that permitted communities to thrive around the Mediterranean basin.

Mediterranean Physical History

By 200 MYA, between the Triassic and Jurassic Periods, Pangaea began moving apart. By 45 MYA, during the middle of the Eocene Epoch, the major pieces of today's continents were reconfiguring as they pulled away from Pangaea. North and South America were separated along a north–south axis. Europe, Africa, Asia and India had also detached. Within that grouping, Europe and Africa, as well as Asia and India, formed a north–south axis. Earlier, India had drifted north from Antarctica and was about to collide with Asia, giving rise to the Himalaya Mountains. The Antarctic continent remained at the South Pole.

Circulating clockwise, the North Atlantic Ocean separated North America from Europe, while the southern Atlantic Ocean, spinning counterclockwise, separated South America from Africa. On the opposite side of the globe, the Pacific Ocean filled the vast distance between the east coast of Asia and the west coast of the Americas. Between Africa, India and Eurasia to the north, and between South and North America, the Tethys Sea flowed east to west, coursing through both the Atlantic and Pacific Oceans, circulating the Earth just north of the equator. (Please see the earlier discussion at the end of the Cretaceous Period).

During the middle of the Miocene Epoch, 15 MYA, the European and Asian tectonic plates were connected and India joined Asia. As a result, the Tethys Sea

T. Sanford, *A Whirlwind History of the Universe and Mankind*,
https://doi.org/10.1007/978-981-97-2674-5_6

was greatly reduced and it now flowed only from the eastern shores of India through today's Persian Gulf (which was still open at the north), passing between Europe and North Africa and westerly into the Atlantic ocean. In the meantime, the Red Sea opened, pushing the Arabian Plate eastward under the Eurasian Plate. This motion caused the Arabian Peninsula to detach from the horn of Africa. In the process, the Arabian Peninsula slid under the south-western rim of the Eurasian Plate, raising the Zagros Mountains at today's Iran-Iraq border. This shift closed the eastern end of the Tethys Sea at today's upper end of the Persian Gulf. As the northeastern arm of the Tethys contracted, it formed the Black Sea, Caspian Sea and Aral Sea across western Asia.

At the end of the Miocene Epoch, 6–5.5 MYA, the northward movement of the African Plate pinched the west end of the Tethys Sea at Gibraltar trapping the Tethys Sea waters and creating the Mediterranean Sea. This sea was partially divided by a south to east ridge of subsurface mountains. The mountains extended from the Atlas Mountains of Morocco in the southwest through Cartage in Tunisia, northeast to Sicily and continuing on to the boot of Italy, forming two connected west to east basins.

The rivers flowing into the Mediterranean, were not enough to keep the basins completely filled. Evaporation in the hot climate depleted the western basin. The eastern half benefited from its depth and being supplied by waters from the Nile and the Black Sea. Eventually, the western half dried completely, temporary exposing massive salt beds on the Mediterranean Sea bottom. This situation did not last long.

By 5.3 MYA at the time of the Miocene-Pliocene boundary, tectonic motion forced the western rim of the basin to subside again. Estimates suggest that in less than two years, the Atlantic Ocean completely refilled both basins, splitting open the Straits of Gibraltar in the process.

Because the African Plate was subducted beneath the Eurasian Plate, the southern shores of the Mediterranean Sea along North Africa were relatively smooth. In contrast, the northern shore was convoluted and mountainous. Here the Eurasian Plate, together with a number of smaller plates squeezed in-between, was pushed up and crumpled by the northward motion of the African Plate into the Eurasian Plate. Plate motions, including that of the Arabian Plate, formed the mountain ranges surrounding the northern and eastern coast line. These included the Alps of Europe and the Taurus mountains of Turkey.

Regional volcanic activity also contorted the landscape. These tectonic energies formed the intricate northern coast of headlands, bays, protected harbors and islands, all of which were ideal for the seafaring civilizations, which emerged once the Mediterranean Sea refilled. Evolution produced the rich volcanic soils upon which future civilizations of Greek, Etruscan and Romans would thrive.

With negligible tides and no surface currents to drive ships off course, the Mediterranean Sea with its high mountains along the northern coast provided landmarks for navigation when far from land and allowed people of differing cultures to connect and trade. In contrast, the undifferentiated southern shore line, aside from the headland at Cartage and the Nile delta, limited seafaring along that coast. The hot dry African coast, abutting the Sahara Desert, was not inviting. These features delayed the development of civilization there relative to that of the northern coast. Despite these limitations, the ease of transporting by sea when compared to land, made the Mediterranean Sea a waterway, linking peoples into developing new dynamic cultures.

Migrations Around the North and Eastern Mediterranean Basin

Recent genetic DNA analyses shows that by 6800 BCE the farmer-herders of central Anatolia began migrating from Anatolia to all areas of Europe. One group traveled along the Aegean Sea (located between the Greek peninsula on the west and Turkey on the east) of the Mediterranean Sea by boat and continued into Greece, then on to the boot of Italy and through Italy to southern France and Spain. Islands, such as Sicily and Sardinia, were colonized by 5200 BCE. A second group pushed through the Balkans along the upper reaches of the Danube to reach northern France and England by 5200 BCE and 4000 BCE, respectively. Later, part of this movement branched and passed through Germany reaching Scandinavia by 4000 BCE. In general, these Neolithic farmers and the original Paleolithic foragers remained separate from each other.

In 6300 BCE, a third group of Neolithic farmers branched from the original 6800 BCE migration out of Anatolia to travel around the northwestern end of the Black Sea. They continued through eastern Ukraine onto the steppes of southern Russia above the Black and Caspian Seas. This region became the homeland of the Yamnaya (3300 BCE), a bronze-age nomadic pastoral people. They derived, in part, from earlier populations on the steppes, which had primarily arrived from Armenia and Iran and likely migrated through the Caucasus isthmus between the Black and Caspian seas.

Prior to the arrival of the Yamnaya, few peoples lived beyond the steppe river valleys. There was too little rain to support farming and livestock. The invention of the wheel used on Yamnaya wagons being pulled by animals changed that. Carrying supplies and water into the open steppes, Yamnayas exploited the vast grasslands. Utilizing the horse, a single Yamnaya was able to herd many more animals than a walking shepherd. In combination, these innovations greatly changed a sedentary pastoral life to a mobile one, leading to a general abandonment of villages. Once on the move, the Yamnaya ventured in all directions.

DNA analyses suggests that after 3400 BCE, the growing egalitarian farming communities of Europe rapidly declined in population for almost a half a millennium. Archeological and DNA evidence supports a Yamnaya westward migration. By 3300 BCE, they were in Romania. Following the Danube, they moved into northeastern Europe by 2900 BCE and into the lower regions of Scandinavia by 2800 BCE. And once on the upper Italian peninsular, they turned westward through the Poe River Valley and entered Spain, France and British Isles. This completed their habitation of the continent by 2200 BCE. In the process, the Yamnaya introduced the horse and wagon and ushered stone-age Europe into the Bronze Age.

Genetic testing of the European population shows how devastating this migration was to the Neolithic farmers. Only half of the DNA genes in today's French, English and Norwegian peoples are that of the Anatolian farmers. The other half is primarily Yamnaya. Along with a little of the earlier hunter-gathers from Africa, the Yamnaya dominated the gene pool.

In general, the Neolithic farmer's genes are greatest in the peoples of southern Europe, with Yamnaya genes being more prevalent in the north. The genes of the Sardinians, being out of the mainstream Yamnaya migration route, are almost 90% Neolithic farmer with only a trace of African and Yamnaya.

The rapid demise of the original farmers may have been caused, in part, by a plague carried by the Yamnaya, which these steppe nomads had lived with for years. The decimation of an indigenous population with no immunity to the diseases of invaders was not unique.

Not only did the Yamnaya migrate westward they also migrated eastward. By 3000–1700 BCE they were in Kazakhstan and the Amu Darya River valleys of Uzbekistan. By 2000–1000 BCE they had crossed into the Indus Valley where they inter-mixed with the local Indians. (Please see previous section: Indus River Valley.) The Yamnaya brought more than their innovations and domesticated livestock. Their spoken speech provided the basis for all Indo-European languages. These included most dialects of Europe, Iran, Armenia, and northern India.

Early City-States in the Eastern Mediterranean Basin

As conflicts among the Assyrians, Hittites and Egyptians were occurring, new states appeared in the lands stretching to the northeastern Mediterranean Sea. These included: Minoans on the islands of Crete and Ancient Thera (2700–1450 BCE), Mycenaeans in the Greek Peloponnesus (1600–1100 BCE) and the Greeks (800 BCE-) throughout modern Greece. Although DNA evidence shows that there was general genetic similarity among these groups, they were not identical. It is believed that modern Greeks descended from these early populations. They all evolved, in part, from the Neolithic farmers that migrated from Anatolia thousands of years prior to the Bronze Age.

The Mycenaeans were first known only through Homer's written epics. It was not until Heinrich Schliemann (1822–1890) excavated Troy in 1870, and the other cities mentioned in the *Iliad*, that they were found to be authentic cities. Similarly, the Minoans were discovered to have existed once Sir Arthur Evans (1851–1941) excavated the 4000-year-old palace of the Minoans at Knossos on Crete in 1900. Both civilizations were created by people indigenous to their regions and not by outsiders, although they were influenced by Mesopotamia and Egypt.

Minoans

The Minoans were considered to be the first advanced European civilization for many reasons. They built four-story complexes, traded throughout the eastern Mediterranean Sea, advanced the technology of shipbuilding, created beautiful artwork and developed a writing system referred to as Linear A (untranslated to date). In addition, they were master bronze craftsmen who used copper mined in Cyprus and grew wealthy from its export.

North of Crete, by 115 km, was the volcanic Island of Thera. Its location was also along the tectonic subduction zone that helped form Cyprus and its easy access to copper. The unstable volcano on Thera erupted (1600–1500 BCE) in one of Earth's most violent explosions. It turned Thera into the cratered island of Santorini, buried Crete in falling rock and ash, while a tsunami destroyed Crete's harbors. This volcano effectively ended trade and starvation ensued.

Mycenaeans

Thera's natural disaster weakened the Minoans and left a power vacuum that was filled by invading Mycenaeans. These were the Greek people of Homer that destroyed Troy in 1184 BCE. They built Mycenae (the city of King Agamemnon who led the expedition against Troy), Tiryns and many other cities cited in the *Iliad*. They constructed the mighty "Cyclopian" stone fortresses on the hills that dominated the farming regions. Critical to the advancement of knowledge was their adaptation of a syllabic script Linear B from the Linear A of the Minoans (1450 BCE). Linear B was a form of Greek known as Mycenaean and was used primarily for accounting materials, goods and keeping records.

Circa 1200 BCE, the Mycenaean civilization began to decline. A probable cause was years of civil discord resulting in the fortress cities being abandoned (1100 BCE). A dark period followed until the rise of the Greek Archaic Period (800–500 BCE) and the Greek Classical Period (500 and 400 BCE). The decline of the Mycenaeans, along with other civilizations of the southeastern Mediterranean Sea during this time, however, has been explained by recent climate measurements. This late Bronze Age crisis coincided with a 300 year drought beginning 1200 BCE causing crop failures and famine. These events hastened a socioeconomic decline throughout the region, which, in turn, precipitated regional migrations at the period's end.

Greeks

For the remaining Greeks, life became simpler after 1200 BCE. The palace life in the fortresses was eliminated and the farmers, herders, carpenters, potters and metal-workers continued their vocations. In 1050 BCE, new pottery techniques evolved that produced beautiful systematical designs. This superior made pottery was referred to as proto-geometric. The process of smelting and working iron was soon developed and by 950 BCE it was used to manufacture weapons and tools, replacing bronze. Although the process of making iron was well known by the Hittites, the Mycenaeans never used the iron ore found in Greece.

During this period, various major settlements like Athens and Corinth, had populations of several thousand. There was a migration of Greeks from the mainland to the coast of Anatolia, where the cities of Miletus and Ephesus were founded. These movements created a Greek presence in eastern Asia, which the Mycenaeans had not.

Greek evolved into a separate branch of the Indo-European language by the second millennium BCE. The oldest surviving words of that language can be traced back to the Mycenaeans in 1300 BCE, when the use of the linear B script disappeared. Circa 900 BCE, it was replaced by writing Greek, using a modified Phoenician alphabet. The Phoenician alphabet was the oldest proto-alphabet known, consisting of 22 consonant letters, with implicit vowel sounds. This semiotic language was derived from the Egyptian hieroglyphs used in the fifteenth century BCE and was spread throughout the Mediterranean Sea by Phonetician merchants.

The key to the Greek written language was the addition of five vowels. This inclusion allowed most of the alphabetical characters to stand for a single spoken sound and was simple enough for anyone to master. A true phonetic alphabet developed and writing quickly proliferated. Scribes, with years of study, were no longer needed to convey knowledge. The innovation of the Greek alphabet with their letter forms, taken in part from the Phoenicians, represented a major breakthrough in the advancement of knowledge.

By 800 BCE, many individual city-states (polis) shaped the core of Greek culture, politics and commerce. The Greeks dominated not only their mainland but the islands of the Aegean Sea and the west coast of Anatolia, referred to as Ionia. The Aegean became a Greek sea. The rugged landscape of Greece, with its steep mountains, deep valleys, convoluted coastlines and islands contributed to the isolation of the various city-states. In contrast to the empires of the Middle East, it was a region of many independent, competing city-states with shifting allegiances that shared a common language and culture—Greek.

This period also saw a rise in religious sanctuaries, oracles and festivals that were pan-hellenistic and which fostered a common heritage. One of the most famous of these events was the athletic contest held every four years at Olympia, beginning in 776 BCE in honor of the God Zeus, and the origin of today's Olympics.

Politically the old chieftain system of government was dead. The independence and isolation of the polis facilitated exploring differing power relationships among the emerging aristocrats, peasants, merchants and artisans. Political systems varied from monarchy, aristocracy and democracy, to rule by a tyrant. Sparta and Athens represented two powerful states with diametrically opposing systems of government. Sparta was ruled by two hereditary kings and a council of elders. In contrast, Athens functioned as a democracy with the Assembly of the Demos, the Council of 500 and the Peoples Court.

Instead of the use of fast chariots as a war weapon in the flat lands of the Middle East, the Greeks, with their convoluted landscape, developed the hoplite formation. For this, foot solders used a shield and spear to fight in a tight phalanx formation so that the shield of one man helped protect his neighbor. These armies employed free men, using their own bronze equipment. The solidarity that developed among the men in the army contributed to the ideals of democracy.

Greece had little arable soil for the cultivation of wheat, and with its growing population, was conscious of its limited food supply. The environment of Greece, however, was ideal for producing olive oil and wine, and good for raising sheep and goats, all of which were traded for grains grown elsewhere. Consequently, Megara, Corinth and Sparta expanded westward, colonizing southwestern Italy and the fertile volcanic soils of Sicily. Naples (700 BCE), Syracuse (734 BCE) and Marseilles (600 BCE) all began as Greek colonies. Athens and her allies pushed eastward establishing colonies along the north shore of the Black Sea in its rich river valleys. Byzantium was founded 657 BCE and later became the future capital of Rome, referred to as Constantinople. These colonies spread Greek civilization through commercial expansion and helped unify the region with their common language.

Phoenicians

Complementing the above states were the Phoenicians (1550–300 BCE), a semitic speaking civilization, which originated in the Levant, during the second millennium BCE. They founded the coastal cities of Byblos, Sidon and Tyre (ancient Canaan). It was during this time that the Jews became an ethnic and religious group that also originated in the Levant, which was referred to as the Land of Israel. Their faith was monotheistic based on a single, all-powerful God.

With the demise of the Minoans, the Phoenicians rapidly filled the trading gap that the Minoans left behind. Using their city harbors on the eastern Mediterranean coast and the cedar forests in the hills of Lebanon, they became expert shipwrights, seamen, and merchants. Between 1200 and 800 BCE, the Phoenicians controlled the Mediterranean trade along the northern African coast. They spread their language and established western colonies to present day Tangier, Morocco. The Phoenicians also settled many western Mediterranean Islands: Malta, a portion of Sicily, Sardinia and the Balearic Islands. In general, the Phoenicians dominated the southern reaches of the Mediterranean Sea leaving the northern shores to the Greeks. Sicily, however, was divided; the Phoenicians occupied the southwestern shore, whereas the Greeks colonized the north and eastern shores.

Using the North Star, the Phoenicians mastered night navigation. They traded as far as Cornwall in England by exiting the Mediterranean Sea through the Straights of Gibraltar and following the coastline north. Tin was brought back from Cornwall, which they smelted with copper from Cyprus to make bronze. They became well known for the purple dye made from the murex snail, used in clothing.

In 529 BCE, the Persian Cyrus the Great conquered Phoenicia. With the decline of Phoenicia, many of the citizens migrated to her westerly colonies, Carthage in particular, in present-day Tunisia. As a result, Carthage became one of the leading trading centers in the western Mediterranean Sea. By 525 BCE, Carthage established complete control over all western maritime settlements, merging into an empire that built fleets of ships and employed armies to advance her interests.

Together, Phoenicia and Greece provided new models of human organization that emerged out of the agrarian civilizations of the east. These two dynamic entities each established numerous city-states that primarily focused on trade. Existing side by side, they split the Mediterranean region into two halves: one characterized by the use of the Phoenician language and culture and the other Greek. Through economic need and commerce they were more innovative than the tribute-taking empires, like the Persians of the Middle East.

Greece, Persia and Hellenization

By 500 BCE, the Persians under Darius I (550–486 BCE) were looking for new territories to annex, and pushed the Ionians into rebellion. The Greeks, in the spirit of pan-Hellenism, formed an alliance between Athens and Sparta to oppose the Persian expansion and triggered Darius to enter Greece in 490 BCE. After decimating Eritrea, which faced the coast of Attica, the Persians continued sailing south where they disembarked at Marathon. There, the powerful Persian army lost to the Athens and a small contingent of Plataeans in 490 BCE. A runner ran to Athens, 42 km (26 miles) away, with news of this great victory before expiring. Today that victory is well known by the modern Marathon run, initiated 1896 in Athens.

Although the war against the Persians was not over, this victory challenged the perception that Greeks might actually beat the Persians. It was considered a pivotal moment in history. To put this achievement into perspective, the Persian Empire had a population of 16 million, where as the Greeks, including both its land mass and Aegean Islands, held only 2.5 million people.

Ten years later, the Persian army under Xerxes, Darius's son, again invaded Greece. This time a limited number of Spartans forestalled the Persians at the Thermopylae pass. Earlier, the Athenians had made the critical decision that the decisive battle would be fought at sea and had built 200 ships to confront the anticipated Persian invasion. It was a strategic decision. After Thermopylae, it was fought in the Straits of Salamis. The Athenian Navy, using superior tactics in the confined space, won. The following year, one more major battle took place, this time at Plataea, led by the Spartans. There, hoplites crushed the bulk of the Persian forces left behind. Greece had stopped Persia from conquering Europe. Over succeeding years, Spartan interests contracted to the coastline of the Peloponnese. In contrast, Athenians expanded their naval victory, liberating Greek cities on the shores of the Aegean and the Sea of Marmara.

The freed Greeks in the Aegean participated in an anti-Persian league, in which they paid fees to the League's treasury (on the island of Delos) based on taxes previously paid to Persia. Eventually the treasury was moved to Athens, at which point Athens effectively became a de-facto maritime empire. From 480 BCE on, Athenian political, economic and cultural growth expanded under the leadership of the elected general Pericles (495–429 BCE). The population of Attica grew to 250,000 people. Athenians entered their Golden Age. This period produced the great philosophers Socrates (470–399 BCE) and Plato (429–347 BCE); historians Herodotus (484–426 BCE) and Thucydides (460–400 BCE); and tragedians Aeschylus (523–456 BCE) and Sophocles (496–406 BCE).

In gratitude for deliverance from the Persians and partly out of the funds extracted from the transferred Delos League, Athens constructed the iconic temple to Athena called the Parthenon. This classic Greek temple, with its elegant colonnades, became a model for future government buildings, capitals, banks and churches throughout the western world.

The gains made by Athens, however, caused the enmity of Sparta. The subsequent 30 year Peloponnesian War (431–404 BCE) destroyed any chance of unity among the Greeks. After a century of infighting, the Greek states were subdued by King Phillip II of Macedonia who reigned from 359 to 336 BCE.

Prior to Phillips assassination in 336 BCE, he and his generals had plans to invade Persia. After his death, the Macedonian crown passed to his 20 year old son, Alexander, who continued the conquest. In 334 BCE the Greeks invaded Persia. Over a series of campaigns that lasted 10 years, the power of Persia, under King Darius III, was defeated and Alexander became the new Emperor of Persia. In 323 BCE, he unexpectedly died at the age of 33 years. Undefeated in battle, he was considered to be one of the most successful military generals in history and was referred to as Alexander the Great. After his death, with no well-defined successors, his generals divided the Empire into four pieces, which they fought over for the next 40 years. By 280 BCE, three kingdoms emerged. Each was ruled by a Macedonian dynasty: the Antigonids in Macedon and Greece, the Ptolemaic in Egypt and the Seleucids in western Asia.

The period between Alexander's death in 323 BCE and the first Roman incursion into Greece, the conquest of the second League of Achaeans in 146 BCE, is referred to as the Hellenistic Period. (The second Achaean League was a grouping of Greek city-states on the Greek Peloponnese that formed in 280 BCE as a buffer to Antigonid, Macedon.) During this time, new Greek cites were established throughout the conquered lands and commercial expansion flourished.

Important among these cities was the port of Alexandria on Egypt's Nile delta. It became renowned not only for commerce but also for its multi-cultural tolerance. It evolved to be a cultural center that competed with Athens. At Alexandria, the wheat surplus of Egypt was sold throughout the Aegean Sea, which brought enormous wealth to the city and furthered the Hellenization of Egypt.

There, Ptolemy II (reigning from 280 to 247 BCE) built the Great Library, which drew scholars from around the known world. Under Ptolemy II, the famous Lighthouse of Alexandria, estimated to be 100 m tall, was constructed and became one of the Seven Wonders of the Ancient World. The Ptolemaic Dynasty was the longest lasting of the original three kingdoms set in place by Alexander's passing and ruled Egypt until the death of Cleopatra VII in 30 BCE. Within 100 years of the passing of Alexander, Greek became the language of not only the ruling class and intellectuals but also of the market place.

Prior to Hellenism, Roman culture had already been influenced by the Greeks and continued once Greece was absorbed into Rome's rule of the Mediterranean. By 100 BCE, Greek culture fully penetrated Roman culture to the point that Roman myths and gods were remodeled into earlier Greek ones. The Hellenized Roman elite further supported the conquered Greeks, making Greek culture indispensable to control of the eastern Mediterranean Sea. Under the subsequent Pax Romana (Roman Peace) Greek culture flourished. This was not true of many of the other civilizations conquered, such as those of Egypt and others in the Middle East. Eventually, Greek culture facilitated by the Romans, would enter European thinking as well.

Evolution of Greek Thinking

Greek Contributions

Greece lay at the northern apex of the Fertile Crescent geographically and was intimately linked through the Aegean Sea to the cultures and ideas that flowed out of Mesopotamia and Egypt through trade and commerce. Building on knowledge absorbed from these exchanges enabled the Greeks to form their own cultural identity that became the bedrock of western civilization. These advances included: architecture, the arts, literature, mathematics, medicine, philosophy and the sciences.

Large scale sculpture was a visual example of this influence. It was from the Egyptians that the Greeks in the Archaic Period learned how to make free-standing sculptures. Eventually, the Greeks transitioned from the rigidly stylized Egyptian forms of 700 BCE, often with Middle Eastern motives like griffins and sphinxes, toward a more natural representation of the human body. By 500 BCE, sculpting evolved to expressing human emotion. Later, the Romans would copy many of these statues in marble. The Discus Thrower, which was originally produced in bronze in 450 BCE, and now has copies in several present-day museums, is a prime example. Its perfection of form expresses the ideals for which the Greeks were known: balance, beauty, harmony and proportion.

Engineering was another example of Greek advances on larger scale. By the 650 BCE, temple architecture progressed forward when limestone and marble replaced wood. Again, the Greeks learned from the Egyptians the engineering skills required to handle the large stone blocks.

Even the religion of Mesopotamia played a role by introducing the idea that creation was the separation of deities from an undifferentiated mass (chaos). From chaos, the deities initiated a series of generational wars until the last generation gained the upper hand and finally brought stability to the universe, resulting in the Greek gods. The Linear B tablets of the Mycenaeans demonstrated that the early Greeks worshipped most of the Mount Olympus gods of the later Greek religion. These included: Zeus, Hera, Poseidon, Demeter, Athena, Apollo, Artemis, Ares, Hephaestus, Aphrodite, Hermes and Dionysus. In the evolution of the Greek gods, the sons of Cronus (time), eventually let Zeus rule the sky, Poseidon rule the sea and Hades rule the underworld. The Earth received no specific god but was ruled by all of them.

Similar to other near Eastern religions, the Greek gods were anthropomorphized: looking, acting, and thinking like humans with all their foibles. But a key feature of the gods was that they were immortal. In contrast to many other religions that promised a wonderful afterlife, like the Egyptians, none was offered immortality except perhaps that ones soul carried on a senseless after life in Hades. No priestly caste existed, as in Mesopotamia or Egypt. Instead, the Greek religious leaders were drawn from the aristocracy. The gods were an integral part of daily life that gave meaning to the world around them. Along with the arts and language, the religion of the Greek gods tied the Greeks together.

Early in the sixth century BCE, philosophical thought began to develop in Ionia, especially in Miletus, where the spirit of exploration was permitted free play. The philosophers were among the first to reject mythical religious explanations for the origins of the universe. Instead they looked for purely physical mechanisms. Thales (624–546 BCE) of Miletus, for example, postulated that the origin of all matter was water, because the liquid transforms into both a solid and gas. The Earth was flat and floated on water. In contrast, Anaximander (610–546 BCE), a student of Thales, proposed that the universe was boundless and consisted of a substance that harmonized opposites such as hot and cold as well as dry and wet. Anaximenes (585–526 BCE), another Milesian, disputed the prevailing theories and postulated that all things were derived from air and that air changed to wind, cloud, condensed into solids or fire when rarefied. Like Thales, he believed that the Earth was flat but, in contrast, that it floated on air.

On Samos, the mathematician Pythagoras (570–500 BCE) was born. Famous for the Pythagorean Theorem, he moved to southern Italy where he taught that arithmetic was key to understanding the universe. Pythagoras proposed that the Earth was spherical and centered in a series of hollow spherical shells. He also thought that the stars were attached to the outer shell with the planets on smaller, interior shells and that movements of these celestial spheres were dictated by well defined mathematical ratios. Pythagoras was the first to differentiate odd and even numbers and conceived the idea of prime numbers.

Critical to the advancement of knowledge, the writings of these Greek philosophers and scientists were widely circulated, discussed and debated throughout the Greek world. Pondering the origins of the universe continued into the next century.

There, Anaxagoras (500–428 BCE) of Ionia, like so many others, was drawn to Athens. Anaxagoras viewed physical material as made of infinitely divisible particles with their organization defined by a force he called intellect. He was the first to give the correct prediction of solar eclipses. He claimed that the moon was of earthly material, the Sun was not a god, but was a white-hot stone, as were all other stars. Another philosopher, Empedocles (493–433 BCE) of Acragas, Sicily, proposed a cosmology based on four primary substances: earth, water, air and fire. He theorized that physical quantities were produced when the twin forces of attraction and repulsion acted upon these substances in different proportions.

Leucippus (460–370 BCE) and his student Democritus of Abdera, Thrace proposed a different view of the universe. They suggested that matter was made of atoms separated by empty space through which they moved; that the atoms themselves were solid, homogeneous and fixed, having different sizes and shapes. They predicted that changes in matter resulted from various combinations of atoms. And whatever was shaping matter was from natural forces unaided by any divine god.

Socrates (470–399 BCE) of Athens developed the basis for western philosophy and logic. Socrates theorized that philosophy should aim for pragmatic outcomes for the benefit of society. Plato (429–347 BCE), one of his students, founded the Academy (387 BCE), where he later taught. Initially, the school provided a forum where his philosophies, along with that of others, could be explained. This school lasted throughout the Hellenistic period as a center of higher learning until 83 BCE, when it was destroyed by the Romans. It was unique for its time in that the school challenged scholars to generate new understandings of the universe. The subjects at the school included mathematics, philosophy and politics.

Hippocrates (460–370 BCE) on the Island of Cos founded the Hippocratic School of Medicine and was considered the father of medicine. Case studies formed the basis of his doctrines. He looked for rational explanations for disease and was credited with advancing the systematic school of clinical medicine. The Hippocratic Oath, *Do no Harm,* is central to all medical training and continues to this day.

The Greek historians Herodotus (484–425 BCE) and Thucydides (460–404 BCE) authored texts that have survived in their complete form. Herodotus of Halicarnassus was the author of *The Histories* and Thucydides of Alimos (located close to Athens) wrote the *History of the Peloponnesian War.* Two quotes from their books, as summarized by S. B. Pomeroy *et al.* in *A Brief History of Ancient Greece* are:

From Herodotus, *he learned from his study of history that power goes to people's, heads, and that the mighty rarely meditate on their condition with sufficient, judiciousness and reflection—that rulers hear what they want to hear, and rush headlong,*
to their own destruction.

About Thucydides, in contrast to Herodotus, who saw *history as an interaction of divine,*
and human forces...Thucydides saw the interactions of people as pretty much exclusively responsible for how things turn out.

Herodotus was often referred to as the father of history whereas Thucydides was referred to as the father of scientific history, due to his evidence gathering and analysis of cause and effect without relying on the Gods.

Hellenistic Contributions

Aristotle (384–322 BCE) of Stagira Macedonia, a student of Plato and teacher of Alexander the Great, founded an institution of scientific learning in Athens, the Lyceum (335 BCE). Like the Academy, it lasted until being destroyed by the Romans in 86 BCE. Aristotle taught at the Lyceum, whose members conducted informal scientific and philosophical explorations. Aristotle was often referred to as the father of science for his logical way of thinking and promoting experimentation wherever possible. Perhaps most importantly, he and the Lyceum promoted the concept of the modern-day scientific method, best encapsulated here from a definition obtained from Oxford Languages: "…a method of procedure consisting in systematic observation, measurement, and experiment, and the formulation, testing, and modification of hypotheses….and where criticism is the backbone of the scientific method."

Euclid (325–265 BCE) of Alexandria, also a student of the Academy, was best known for the way space and shape were perceived and for his famous text book, *The Elements of Geometry,* where geometry was presented in axiomatic form referred to as Euclidean geometry. Euclid was the first person to write a complete review of geometry and was considered to be the father of geometry.

Eratosthenes (276–194 BCE) of Cyrene (today a coastal city in Libya) was best known for calculating the circumference of the Earth. Eratosthenes used the measured shadow cast by a vertical pointer at Alexandria and directly overhead at Syene (now Aswan) during the summer solstice. Together with the known distance between the two cities plus a little trigonometry relating to triangles he arrived at 48,000 km. This estimate was remarkably close to the present value of 41,000 km. He was a mathematician, geographer and astronomer who was educated at the Academy and later became the chief librarian at the Library of Alexandria. He was also the first to estimate the tilt of the Earth's axis and to create the first global projection of the Earth using meridians and parallels, thereby establishing scientific cartography. Today, Eratosthenes is referred to as the father of geography.

Archimedes (287–212 BCE), of Syracuse, Sicily, was a polymath: mathematician, physicist, engineer and inventor, who also spent time in Alexandria. Archimedes is famous for discovering the Law of Hydrostatics, which states that a body not fully immersed in a fluid weighs the same as the amount of fluid it displaces. In the field of mathematics, using techniques of infinitesimals, he was able to calculate the value of pi, the ratio of the circumference to diameter of a circle, and to prove a range of geometrical theorems, including the surface area of a circle and sphere and its volume. He introduced the concept of the center-of-gravity to the study of physics. In the field of engineering, Archimedes discovered the laws of pulleys and levers that allowed one to lift heavy objects, using reduced forces. The water screw was just one example of his many inventions.

The work of Claudius Ptolemy (100–170 AD) of Alexandria was known to every early European mariner. He was a mathematician, astronomer and geographer who considered the Earth was at the center of the universe and is famous for his Ptolemaic system. His astronomical work was enshrined in 13 books contained in the *Almagest*. He was able to show in his first book, by rational arguments, that the Earth was at the center and that no contrary argument could be found to dispute that claim. As a result, Ptolemy's system became dogmatically embedded in western Christian theology until the fifteenth century.

As a geographer, Ptolemy's reputation rests on his *Guide to Geography*, which was partitioned into eight books. It included information on Europe, Africa and Asia, tabulated according to longitude and latitude. Unfortunately, it contained a few errors, such as estimating the Earth's circumference to be 30% less than it actually was. This miscalculation had ramifications centuries later when Christopher Columbus, in search of an ocean passage to Asia in 1492, underestimated that distance.

The evolution of the philosophers, mathematicians, scientists and intellectuals, together with the establishment of centers of higher learning, represented new steps in human advancement. This development was made possible, due to the increased contact and communication among cultures that was facilitated by trade and the economic activities that promoted growth.

Rome: The Evolutionary State in the Western Mediterranean Basin

Advances in human social and cultural development, along with Greece's limited scientific and technological knowledge, gave humanity a preview of how the universe may have evolved. Another pivotal step was the development of the Roman Empire, which absorbed much of the Greek culture. Rome's forceful incorporation of Western Europe into its sphere of influence imprinted its written language, way of thinking and political organization on Europe for the next five hundred years. It was in Europe, after another half millennia of social and economic progress, that extensive advances in scientific understanding were made. The focus here, accordingly, moves to Rome and later to the West.

Rome's Beginnings

As the eastern Mediterranean empires grew, the western Mediterranean was developing into regions of cultivation. The most important of these areas occurred on the Italian peninsula. It was there that Rome formed near its western center. Rome was located 30 km inland from the Tyrrhenian Sea on the banks of the Tiber River in a grouping of seven hills above the plains of Latium. Rome was founded by the legendary date of 753 BCE. At this time, the peninsula contained a conglomeration of many different Italic tribes, Greek colonies in the south and the urban Etruscans of Tuscany in the north. (Herodotus claims the Etruscans came from Lydia in western Anatolia.) Not only was the peninsula fractioned politically but linguistically as well. With the exception of the Greeks and Etruscans, the people of the peninsula spoke a prehistoric Indo-European language. Their alphabet reflected Greek and Phoenician scripts, which produced an early Italic alphabet.

The Greeks to the south, together with the cultured Etruscans to the north, significantly influenced the urbanization of the Romans. From the Greeks, they acquired religion, literature and the basics of architecture. From the Etruscans, the Romans adapted a drainage system and pursued gladiatorial combat for entertainment. During its start, the Romans were borrowers and open to the ideas of others. In general, they improved upon the concepts and techniques they adopted from others.

Early in its growth, Rome had been coveted and ruled by Etruscan kings. Over time, the Etruscans were weakened by the intrusion of Gallic tribes from the north and succumbed to Roman forces in 509 BCE. At that moment, a newly formed Roman Republic was never again to be ruled by a king.

In the new republic, the highest government positions were held by two Consuls (leaders). A Senate consisting of aristocrats elected the Consuls for one year terms. The Senate held the real power and was the equivalent of the Greek Council of Elders, which decided policy. The presence of this powerful body of peers held individual aristocrats in check. Both Consuls and policies required confirmation by the peoples' vote. Decrees were issued in the name of both the Senate and the people.

Socially, the population was split between two classes: the commoners (plebeians) and the aristocrats (patricians). A cultural development occurred in 450 BCE that helped stabilize the state. This was a set of laws that applied to everybody, plebeians and patricians alike and were inscribed on 12 bronze tablets. They were posted in the forum where they could be seen by all. Plebeians read them and become familiar with the contents, which helped reduce the abuse of power by the patricians.

Continued conflict between the plebeians and patricians for a more equable distribution of power led to many changes over the Republic's long history. By 300 BCE, the plebeians gained the right to elect their own representatives (the Tribunes), who had the power to veto consular decisions. These changes helped contribute to the stability of the Republic. Over time, the laws were modified and replaced by ones more germane to the emerging society. Eventually, they formed the foundation for what became a system of codified law that governed the world of the Romans. With time, Roman law would become a model for the formation of many governments, including some in the present day.

During Rome's beginnings, a number of city-states like Rome, had grown up within Latium, however, Rome became one of the largest. Its size was assisted by their early Etruscans kings who expanded their territory. Driven by external threats, these city-states formed coalitions for defensive purposes. Eventually, Rome assumed military leadership over its alliances and exercised guidance without eliminating self-rule. After the military campaigns, the spoils of war were shared among participating alliances. This scheme became the prototype for future Roman expansion.

The invasion and sacking of Rome by the Gauls in 390 BCE was not fatal. It motivated the Romans to enhance their military and promoted their expansionist policy. During the fourth century BCE, Roman control increased. Newly subordinated communities were again allowed to maintain local control with the proviso that they supply military manpower for the Roman legions. In this fashion, alliances were established with individual polis bound by bilateral treaties throughout the Italian peninsula. This strategy allowed Rome to coordinate military activity on a larger scale without having to contribute to increased spending on non-Roman civic arrangements.

Rome deliberately decided that granting the status of citizenship to outsiders and defeated enemies was a worthy political strategy. Divorcing citizenship from physical location or ethnicity built a more homogenous sense of cohesion. A similar policy was applied to the Roman military where intermingling of soldiers of differing backgrounds broke their allegiance to a given or defeated community. Extended military service furthered the sense of being Roman.

The Roman state became primarily focused on warfare with religion and war being closely aligned. As the Roman State established control over Italy, temple building in Rome expanded. Allied polities were protected by Rome and a sense of security prevailed. The Roman State and its alliances grew from 50,000 in 400 BCE to more than three million by 260 BCE. With the entire peninsular under its influence, the Romans were able to field armies as large as 40,000 men, rivaling those of the neighboring Macedonian Kingdom. Eventually, Rome was able to control the entire peninsula without significant interference from such powerful empires to the east.

Rome's Expansion Beyond Italy

Once Italy was subdued, Rome looked to Sicily for resources. At this time Sicily was dominated by both Carthaginians and Syracusians. At the start of the conflict, the Syracusians joined the Romans side, in what is referred to as the first Punic War (264–242 BCE). The Carthaginians were the major naval power in the western Mediterranean Sea and Rome was primarily an agrarian society. To be competitive, Rome reengineered captured Carthaginian war ships and rapidly constructed a competent navy. She improved the ships design, by adding a naval boarding device, which permitted Roman soldiers to board an enemy vessel quickly. Eventually, the Romans prevailed in subsequent naval engagements, thus ending Carthaginian domination of the sea. Sicily, as well as Sardinia and Corsica came under Roman jurisdiction. To administer the new territories, Rome adopted the provincial system used by the Persians. Each new territory was controlled by a governor selected by the Senate and called a proconsul.

The conflict with Carthage lingered causing a second Punic War (218–201 BCE) when the Carthaginian general, Hannibal (247–183), crossed the Alps and invaded Italy. For years, he successfully out maneuvered every Roman consul sent to curtail his destructive path through Italy. An existential battle with Hannibal occurred at Cannae, Apulia, in 216 BCE. In a single day, Hannibal, with a smaller army, decimated the largest army ever amassed by Rome and over 80,000 Romans were killed. The Carthaginian success is considered to be one of the greatest military victories in history.

Rome, however, was not finished. Rome continued to raise legions of men through her alliances but never again directly engaged Hannibal in Italy. In 204 BCE, Publius Scipio, authorized by the Senate, took the war to Carthage and defeated its defending army. Hannibal was called home and was also defeated when Carthage surrendered to Scipio in 202 BCE. This long war illustrated the power of Rome's strategy of military alliances. Each ally continued to contribute men to the war effort, allowing Roman rule and the Senate to survive. Rome's steadfastness of its people and the Senate, however, had been severely tested.

The importance of the war and the peace with Carthage established Rome as the master of the entire western Mediterranean Sea. This conflict split the classical world in half: the western half under Roman domination and the eastern half under the residual conflicted Hellenistic kingdoms of Alexander's Empire. These were the Macedonians, Seleucids and Ptolemaics.

The eastern half had not been entirely neutral during the conflict. King Philip V of Macedonia had allied with Hannibal and King Antiochus III of the Seleucids provided a safe haven for Hannibal after the war. These considerations provided a pretext for Rome to enter Greece in 198 BCE. Philip's army was defeated and pushed north to Macedonia. Greece was declared free by the Senate and made a protectorate of Rome. The people of northern Greece, however, were not content with Roman rule and invited Antiochus to deliver them from the Romans. The Romans quickly subdued Antiochus, crossed into Asia after 190 BCE, requiring him to surrender Anatolia. By this time, portions of the Seleucid Kingdom in Anatolia were fracturing into individual sub-kingdoms like Pergamum, facing the western Aegean in Anatolia, or Pontus, facing the Black Sea in eastern Anatolia.

In the western half, 151 BCE, Carthage went to war with the Numidians, breaking conditions of a previous treaty with Rome. Rome exercised a legalistic pretext to initiate war with Carthage and finally destroyed the city in the third and final Punic War (149–146 BCE), thereby gaining control of Numidia. The continuing inevitability of Roman expansion in the Mediterranean basin led the last King of Pergamum, with no heir, to will his Kingdom to the Romans in 133 BCE as did Cyrene in 96 BCE.

In 67 BCE, the Senate authorized Pompey (106–48 BCE) to eliminate the pirates from the Mediterranean Sea, and later to suppress Mithridates VI of Pontus. By removing the pirates, the grain shipments feeding Rome were secured. By 64 BCE, the entire coastline from Pontus (southern coast of the Black Sea on the northern coast of Anatolia) to the borders of Egypt was brought under Roman control with the interior kingdoms becoming Roman vassal states. With the exception of the Ptolemaic Kingdom of Egypt, the Romans commanded the entire Mediterranean Sea. In 48 BCE, Julius Caesar (100–44 BCE) entered Egypt and a few years later Egypt also succumbed to Roman control.

In summary, Rome's steady expansion around the Mediterranean Sea was the result of a number of factors. Foremost was the ability to mobilize its extended military. Second, was the political structure of republicanism. Another contributing factor was the relative isolation of the Italian peninsula, which was protected by the Alps to the north, making border protection costs minimal. With little interference from the east and command of the Mediterranean Sea, military movement and food supplies were assured. A more flexible mode of fighting with a superior choice of weapons further contributed to Rome's success. Most importantly, in contrast to other regional powers, Rome granted the colonists, loyal allies and even freed slaves, Roman citizenship. This liberal growth of the citizen body meant Rome always had man power to meet its military needs as it expanded. All of this fostered a sense of loyalty to Rome.

End of the Republic

The Roman Republic, despite its many favorable policies, did not end well. Threats to the Senate and civil wars led by competing generals plagued the last century of the Republic (100–27 BCE). The incredible wealth and number of slaves that began to flow into Rome after the fall of Carthage and the kingdoms of the east upset the relationship between the commoners and the aristocrats.

Notable was Marius (157–86 BCE), initially a plebeian tribune, who had the support of the people and installed a popular government. Later, the patrician Sulla (138–79 BCE) restored Senatorial rule. Central, during the following period, were Pompey, the ablest lieutenant of Sulla, and two other politicians were Julius Caesar and Crassus (115–53 BCE). Caesar controlled what remained of the Marius Popular Party and Crassus was the richest man in Rome. The three—Pompey, Caesar and Crassus—formed an informal secret alliance. They created a Triumvirate (60–53 BCE) whereby each would use their respective influence to help one another achieve their individual goals, pushing aside the rule of the Senate.

They decided that Pompey would rule in Rome, Caesar would take military command in Gaul and Crassus in the Middle East. Caesar, in a series of military campaigns between 58 and 51 BCE, subdued Gaul and established the Roman frontier on the Rhine River. From there, he surveyed the Island of Britain. Crassus was killed, during an ill-conceived confrontation with the Parthians (r 247 BCE-224 CE) who had taken over much of the old Seleucid Kingdom. Caesar and Pompey were left to confront one another.

The success of Caesar in Gaul earned him the devotion of his army but raised the concern of the Senate. The Senate turned Pompey against Caesar by convincing Pompey that it was his duty to protect the constitution. Caesar was instructed by the Senate to disband his army before returning from Gaul. Instead, fearing for his life, he crossed the Rubicon River in 49 BCE (a metaphor now referred to as a point of no return) and entered Rome, precipitating a civil war with Pompey. Eventually Pompey's forces were neutralized and Caesar became dictator for life. On the Ides of March, 44 BCE, Caesar was assassinated by a number of Roman senators who feared that he wanted to become king.

Early Roman Empire

A long series of civil wars ensued that ultimately ended the Roman Republic and initiated the Roman Empire under Augustus 27 BCE. During the transformation, Caesar's appointed heir, Octavian (63 BCE-14 CE), defeated the army of Caesar's general, Mark Anthony, at the battle of Actium, Greece (31 BCE). There, Mark Anthony fled with Cleopatra to Alexandria, Egypt where they committed suicide. Ultimately, Octavian, unchallenged, became emperor taking the name Augustus, meaning "lofty one."

To appease the Senate, Augustus founded the principate, a monarchy controlled by an emperor (himself) having power for life with his powers hidden behind constitutional reforms. He worked to placate the Senatorial class and to encourage its participation in the new hierarchy by establishing many of the traditional offices. He governed the more difficult provinces while allowing the remainder to be controlled by the Senate.

Augustus gave Rome the first sense of a stable government after nearly a century of discord. The peace and prosperity that followed was known as the Pax Romana. Under Augustus, the Roman frontier in Europe was extended to the Danube River (14–8 BCE). Throughout his reign, many of the earlier Roman vassal states were annexed, Egypt and Cyprus in 30 BCE and Judaea in 6 CE. Earlier, Caesar had reestablished Carthage as a Roman colony and Augustus continued its redevelopment, enabling it to become one of Rome's important colonies. The bulk of Rome's 28 legions (a legion consisted of 5000 men) were stationed along frontiers. Only the Praetorian Gard, which protected the Emperor, was garrisoned in Italy.

During the Pax Romana under Augustus, the new empire and particularly Rome flourished. Rome expanded to a population of 250,000, more than twice the size of Alexandria's 90,000, which was the earlier world record holder. By contrast, Athens had a population of 30,000.

Rome's needs for water were satisfied by the construction of numerous aqueducts that spread across the area of Campagna from the surrounding hills and mountains, however, wheat had to be imported from the provinces. At first, wheat from neighboring Sicily, Sardinia and North Africa was adequate. By the end of the Republic, however, these sources were no longer sufficient. Wheat from Egypt bridged the short comings and the Mediterranean Sea trading network was rearranged for the benefit of Rome.

Beyond the Mediterranean Sea other important changes were occurring. Contacts between China and India were being establish by trade. Caravans formed routes through central Asia, while the sea routes through the Red Sea and across the Arabian Sea, were used to import trade items like silk from China and spices such as pepper from India.

The governmental architecture Augustus established continued to function relatively smoothly until Nero (37–68 CE). In 68 CE, the governor of Spain declared Nero unfit and successfully attacked Rome. Nero died by his own hand and the Roman Empire dissolved into a series of civil wars with no clear winner. In 69 CE, General Vespasian (9–79 CE) restored order and by 75 CE, had expanded the Roman empire to England and Wales.

On August 24, 79 CE, Mount Vesuvius erupted destroying Pompeii and Herculaneum along with many towns along the Bay of Naples. These geological events would occur throughout the rest of Italian history. Unbeknownst to the Romans, Italy resides over fault lines generated by the northward compression between the African and Eurasian tectonic plates.

Limited expansion continued under subsequent emperors, with Trajan (53–117 CE) adding Dacian (north-central and western Romania) to Rome's first and only trans-Danube province. By this time, Rome had expanded into the low lands of Scotland, Armenia and Mesopotamia. Hadrian (76–138 CE), Trajan's ward and successor, felt the Roman Empire was big enough and pulled the boundary back to the Euphrates. In the process, Mesopotamia was lost and Armenia was returned to being a client kingdom. In Britain, Hadrian built his wall across the width of Britain, separating the barbarians from the Romans, and effectively abandoning Scotland.

Following Hadrian, the succession passed successfully to Antoninus Pius (86–161 CE) and subsequently to Marcus Aurelius Antoninus Augustus (121–180 CE). Traditionally, each emperor picked his successor based on skills. This was not the case, however, when Marcus Aurelius chose his incompetent son, Commodius (161–192 CE), to follow him. Like the death of Nero, his assignation generated a civil war. Candidates from each of the eastern armies vied to be emperor. The commander on the Danube River who fielded the largest army, Septimius Severus (145–210 CE) eventually prevailed, ruling from 193 to 211 CE.

Roman Empire's Golden Years

By the end of Marcus Aurelius' rule, Rome had built a vast urbanized empire that spanned the Mediterranean region linking the shores of three continents: Europe, Asia and Africa. Rome ruled through their cities, the associated aristocratic families and their interconnections. Not only was the Mediterranean Sea now filled with multiple shipping lanes with safe harbors but the bordering lands were further inter-connected by a paved road system that spanned over 80,000 km. These systems of infrastructure advanced communication, commerce and the movement of Rome's armies.

Rome had molded an enormous area of different ethnic peoples militarily into a common empire with a homogenous cultural entity by eliminating tolls and trade tariffs and by establishing common laws and language. At this time, Rome controlled approximately 75 million people over an area of 6.5 million square kms, with the city of Rome consisting of a million people.

Peace within the empire was protected by 30 legions stationed clockwise around the regions of its perimeter: 3 in Britain, 4 along the Rhine, 10 along the Danube, 8 in Anatolia and the Euphrates, 1 in Egypt, 1 in Numidia, 1 in Spain and 2 in the Alps. The peace generated a bounty shared by all. Although Rome was at the center of all policies and economic activities, Rome was not top heavy. Many other cities expanded and developed as well. These included: Alexandria, Antioch and Carthage with 200,000 people each. Many Hellenistic cities to the east expanded beyond their initial size, such as Pergamum that reached a population of 120,000.

The sustained peace, law, enforcement of property rights and a uniform currency motivated increased trade and the associated development of technology. The law decreased transaction costs, while trade increased specialization and technology promoted new tools. This enabled laborers to extract more energy for productive applications. The increased movement of goods and people opened new markets and opportunities for the growth of a middle class and social development. Progress in Gaul introduced metal tools and better plows. Improved screw presses, new machines to lift water and water mills evolved. Techniques for mass producing goods developed. Metallurgy and mining techniques improved along with transport technology. Ships became faster and larger, while the improved port facilities increased safety. The sheer size of the empire amplified these continued improvements.

Although the wealthy class received the bulk of the benefits of Roman growth, the lower classes also benefited in many ways. Despite the influx of slaves, the higher wages earned by unskilled laborers illustrate the power of the Roman economy. Wage improvement of the lowest class of laborers stayed ahead of increasing rents and prices. One in five of the empire's population lived in urbanization. Such a level of social organization would not have been possible without significant economic advances over former eras; in prior civilizations, it took nine farmers to support one city person.

A massive building boom illustrated the spread of wealth throughout the empire. Rome was one of the biggest beneficiaries. In 400 CE, Rome still had 37 gates, 19 aqueducts, 1352 cisterns, 856 baths, 144 public latrines, 290 granaries, 254 bakeries, 28 libraries, 2 circuses and much more. These unique features of Rome were spread throughout the empire. They included colonnades with porticos and arches, amphitheaters, public baths, a forum, temples, shrines, basilicas; and other amenities such as latrines, sewers, running water, drains and heating systems. Some of these structures outside of Rome were marvels of civil engineering such as the great aqueduct that brought water to Carthage. It conveyed water more than 120 km and was the longest water supply structure ever built by the Romans.

Today, many remnants of these constructions are still visible. In contrast, Roman temples to Roman gods are nearly absent or were converted to churches. Originally, Roman temples were among the most important buildings. Their form was greatly influenced by the Greek temples. Their construction and maintenance was a major part of the Roman religion. All towns of any significance had at least one. Romans were religious with its religion of polytheism being decentralized. The priesthood and temples were loosely embedded in the lives of the empire's towns and villages. In general, Romans incorporated the gods of the cultures they conquered, thus expanding their gods.

Within Rome, two buildings illustrate the advancement of Roman civil engineering: the Colosseum and the Pantheon. The Colosseum was an oval shaped amphitheater, initiated in 72 CE and completed in 80 CE, that held 50,000–80,000 people. Just the outer wall, which was 48 m high, was estimated to be made of 100,000 cubic meters of travertine stone. The numerous stone blocks were positioned without mortar, using just 300 tons of iron clamps to secure their placement.

In contrast, the Pantheon was a unique Roman temple dedicated to all the gods of Rome. The original Pantheon was built in 27 BCE and burned down repetitively. Reconstructed out of stone and Roman concrete, it was rededicated in 126 CE. The structure was cylindrical with a dome and fronted with a portico. The dome is still the largest unreinforced concrete dome in the world. The enormous dome, originally covered in bronze was supported by the lower walls of the rotunda, which were six m thick. The portico was held high by large granite Corinthian columns. Each column was 12 m tall, 1.5 m in diameter and weighed 60 tons. The columns were quarried in Egypt, brought by ship across the Mediterranean Sea and transported up the Tiber River. The ability to transport the columns over such long distances illustrated the sophisticated Roman organization and engineering abilities of the period.

In summary, the early years of the empire represented real growth on a per capita basis. The borders were secure, the civil administration was stable and the citizens were free to engage in creative, peaceful activities. The massive scale of assembling the many different polities, each with its own culture, history and language into one empire was singular. Today, that empire would require the coordination of 32 separate governments. The time between 96 and 180 CE was judged by the historian Edward Gibbons (1737–1794) as being the "most happy and prosperous." Rome's evolution represented a new level in human development.

Roman Empire's Declining Years

The expansion of Rome prior to Marcus Aurelius's rule occurred during what has been identified as the RCO (Roman Climate Optimum). It was a period of stable, warm and wet weather throughout most of the Roman Empire. The climate helped provide the needed conditions for agricultural growth and extended farming into new areas of cultivation, in particular to those at higher elevations. The RCO helped turn the lands ruled by Rome into more land suitable for cultivation than any time before or after. Fertile agriculture was a key engine along with trade and improved technology, which helped to drive the expansion of Rome. These conditions, however, were about to change slowly over the next three centuries.

Over these centuries, the Empire's fortunes became a complex interplay among weather, pandemics and imperial restructuring. Following the RCO, climate variation appeared in the later part of the second century. Episodes of Rome's decline then followed from pandemics, the beginning of Christianity, the transfer of the empire from Rome to Constantinople and the loss of the Western Empire.

The Antonine Plague (166 CE)

By mid 166 CE, the Antonine Plague arrived in Rome, so named by the famous Roman doctor Claudius Galen (born Pergamon 129 CE, died Rome 216 CE). Prior to the pandemic, he was brought to Rome by Marcus Aurelius. Galen promoted the ideas of Hippocrates who put great focus on clinical observation and anatomy. He was a prodigious author of books, believing that knowledge should be shared. He had studied in the famous medical school of Alexandria, Egypt. Galen's work dominated medicine in Europe and the Arab world for the next 1500 years.

The plague was probably caused by smallpox that originated in Africa or the far east and was transmitted to the west through the sea lanes connecting the Indian Ocean to the Red Sea and to Alexandria. From Egypt, it moved from the southeast to the northwest through the Roman shipping and road networks, eventually reaching Britain. The dense Roman urbanization was ideal for spreading this infectious disease. Modern estimates gauge the mortality from the pandemic to be 10% of the Roman empire. Although the plague produced a large loss in population, depleting the armies protecting borders and the mining of silver for the money supply, the Imperial order had enough reserves to withstand the shock. Politically capable provincials filled the gaps in the Imperial system and moved up in stature. This transition was a mini turning point in Roman society.

The Roman Empire revived under Septimius Severus (145–211 CE) who brought many Roman colonists from its vast territories into the political order of the Senate, including Egyptians. In 212 CE, his son Caracalla (188–211 CE) granted full citizenship to all free inhabitants of the empire, affirming that Rome was indeed a territorial state. Caracalla continued Septimius Severus' constructions by building monumental baths on the edge of Rome. An additional aqueduct was constructed by Alexander Severus (208–235 CE). Large water mills and giant granaries ringed the city. The building boom illustrated a demographic recovery. In 248 CE, the city of Rome celebrated its thousandth year birthday. The unwalled city filled the terrain between the hills and exemplified the Pax Romana.

The Cyprian Plague (250–270 CE)

Prior to 249 CE, the Cyprian Plague had arrived in Ethiopia and migrated north and west throughout the Roman empire. By 249 CE it was in Alexandria and in 251 CE it was in Rome. It spread rapidly for nearly 20 years. At its height, an estimated 5000 people a day died in Rome. In Alexandria, as many as 62% of the population perished. The most likely cause of the plague was from a family of hemorrhagic viruses: filoviruses represented by the Ebola virus.

A global turbulence in the climate had affected the monsoon seasons, precipitating the ecological changes that influenced the plague. In the 240 s CE, a destabilizing drought had afflicted the southern region of the Mediterranean basin. In contrast to the Antonine Plague, which relied on the food reserves of the Empire, the Cyprian Plague was more devastating, as there were few reserves of food remaining.

The Roman army was depleted by the pandemic and the frontiers collapsed by early 250 CE. In that year, the Goths invaded the Danubian front and in 252 CE the Romans lost control of the entire Danube line. The Rhine system collapsed in mid 250 CE. In 256 CE, the Alemanni and Franks invaded the provinces of Gaul and in 260 CE, an incursion from the upper Danube reached Rome. In the east, the frontier on the Euphrates fell. The Persians (a new Persian dynasty that had over thrown the Parthians in 224 CE) overran Asia Minor and Syria. Between 248 CE and the rule of soldier-emperor Claudius II (214–270 CE) in 268 CE, Rome was affected with failures: taxes and currency were impossible to control and an accelerating disorder consumed the empire.

Once the drought, plague, destruction and the monetary declined, a new imperial system coalesced under Claudius based on two principles: the emperors would be part of the Danubian military and their army would be paid in gold. This military autocracy was in contrast to the former rule by limited senatorial and city elites. Like previous emperors, however, these emperors were Romans first who showed strong support for Roman Law and were dedicated to protecting the empire as a whole. The immediate successor to Claudius, who was lost to the plague, was Aurelian (214–275 CE). He dedicated his talents to the reconquest of the lost provinces, had walls extended around the city of Rome between 271 and 275 CE and reformed the currency.

Rise of Christianity

Throughout this period, the inadequacy of the pagan religions (gods) to stem the tide of the plague became apparent and hastened the progress of Christianity. That religion had been incubating ever since Saint Paul (5–65 CE) began preaching. Prior to the end of the second century, Christianity could perhaps account for only 100,000 believers. In contrast, the third century began with an acceleration of adherents to Christianity despite Christians being persecuted. The Christian ethic of love, compassion, respect for the individual and comfort for the sick was a great inducement into the faith. The promise of resurrection removed the fear of death and greatly encouraged believers.

The transition towards Christianity was perhaps one of the most significant changes that resulted from the devastation of the plague.

In contrast to paganism's local organization, Christianity was disciplined and an ordered structure from the beginning. With time it evolved into a mirror image of the civil administration of the empire itself. The office of the bishop, an invention of the late first century, ruled over a diocese similar to the secular provincial administrative districts of the Roman Empire.

Following Aurelian, a series of five short-lived emperors ruled until Diocletian (244–316 CE), a Danubian general, brought stability and efficiency to governing in 284 CE. He concluded that the empire was too big to be run by one individual and divided it in half a year later: into western and eastern sections, with each emperor selecting a Caesar to follow when the emperor passed. This tetrarchy system was short lived. Constantine (272–337 CE) subverted the system by becoming the sole emperor in 324 CE.

Constantine credited the god of the Christians for his critical victory at the Battle of the Milvian Bridge, Rome, in 312 CE, where he experienced a vision of a cross. Christianity presented Constantine with a mechanism for uniting the empire with a common agenda. His rule was transformational in a myriad of ways. He moved Roman support for pagan gods to one Christian god, legalized Christianity and he moved the capital from Rome to new Rome (Constantinople) in 330 CE. Resources went from pagan temple building to church constructions, such as St. Peters in Rome and the Temple of the Holy Sepulcher in Jerusalem. At this point, the Church and State began to evolve into a new society: Christendom.

Constantine generated a second senate for his new capital, one that created new mechanisms to achieving senatorial status. This inflated Senate grew, based on reward for imperial service. Resources, like the grain from Egypt and new construction, consequently went to Constantinople. Rome, however, still remained the symbolic center of the empire, retaining great wealth and a fourth century population of 400,000. Importantly for Rome, the senior bishops of the other major centers of Christianity, like those of Alexandria and Antioch, recognized the superiority of the Bishop of Rome over that of the other bishops. Eventually, the Bishops of Rome would become the popes of the future. The Roman congregation, after all, was founded by St. Peter (1–64 CE) himself.

The move of the capital to Constantinople was a brilliant strategic move. Land locked Rome was no longer at the center of commercial and military activity. Constantinople was, being both near the Danubian frontier and the wealthy eastern provinces. Its position on a triangular spit of land, jutting into a commanding position on the Bosporus Strait, was easily defensible, especially once a defensive wall was constructed at its western end between the Sea of Marmara and the Golden Horn.

Under Constantine and his sons, gold became the basis of the new economy. Being allowed to float on a common market, gold revived the financial sector. Commerce returned to the Mediterranean Sea with Constantinople now at the corner of a giant free-trade zone. Urban life rebounded throughout the empire. Constantinople itself in less than 100 years expanded from a population of 30,000 to 300,000.

Until the time of Justinian (482–565 CE), growth in monumental engineering structures, like the mammoth cistern for storage of water or aqueducts feeding water to the city, continued. For protection (motivated by the sack of Rome in 410 CE), a second outer wall 5.5 km long was constructed as a double wall with moat. This wall, built from 412 to 414 CE, enlarged the territory controlled by the expanding city. Its effective design, with a 12 m high inner wall imbedded with 96 massive towers every 53 m, protected the city until it was breached by the Ottomans in 1453 CE.

Climate Change in the East and the End of the Western Roman Empire

The Roman Imperial system had been centralized between Diocletian and Theodosius (347–395 CE) and became more ritualistic. At this time, the Roman Empire was the most powerful state in the world and had all appearances of being stable well into the future. Unbeknownst to the west, however, brewing on the steppes of Eurasia was an extreme aridity that would change the trajectory of Roman history forever. On these grasslands, the cooling caused droughts, reducing the resources for feeding flocks and horses of the Huns and therefore initiated migrations of the Huns westward. By 370 BCE, they reached the Don River forcing other nomads and farmers off of their land.

For over 100 years, the territories north and east of the Danube River had been settled by Goths in relative compatibility, within, the constraints of the Roman frontier. In 376 CE, however, more than 100,000 Goths pushed up against the frontier seeking asylum within its interior from the advancing Huns. Eventually the Romans yielded and allocated them lands along the Danube frontier. Poor treatment of the Goths resulted in open revolt. The eastern army, sent to put down the rebellion, was annihilated near Adrianople (today Edirne, in northwest Turkey) in 378 CE. Now weakened and not able to hold future migrations back, the Romans under Theodosius initiated a policy of allowing settlement of entire peoples on Roman soil. They were, however, obliged to provide service in the Roman military under their own leaders, making them allies.

Theodosius, was the last emperor to rule both halves of the empire. During his reign, Christianity became the only religion allowed. After his death, his two sons inherited the empire, with Honorius ruling the western half from 395–423 CE and Arcadius ruling the eastern half from 395–408 CE. Court intrigue between Rome and Constantinople eroded the empire's response to the continuing frontier crises and crossings. In 405 CE, a new migration of Goths crossed Noricum (Austria) and plundered Italy. In the winter of 406 CE, another contingent of barbarians, including Alans, Suevi and Vandals crossed the frozen Rhine, pillaged Gaul and pushed into Spain. These barbarians would never be removed. Eventually the Vandals crossed into North Africa. Roman control of the territories beyond the Alps, especially Spain, northern Gaul and Britain, disappeared.

In 395 CE, Alaric (370–410) unified the Goths who had settled in 382 CE. Seeking better arrangements from the western Empire, in 408 CE, he led his warriors across the Alps and surrounded Rome. The ancient capital was held hostage for a ransom, which Honorius (who ruled from Ravenna) refused to pay for three years, before Alaric's forces entered Rome on August 24, 410 CE and sacked the city before leaving. Subsequently, the piecemeal loss of territory and control continued. Barbarians moved into the western Empire: Ostrogoths into Italy, Burgundians into Savoy and Visigoths into Aquitaine. Although the native Romans outnumbered the barbarians, the barbarians eventually controlled the machinery of the state. At this point, the western Empire lost its ability to coordinate its military.

In addition to the westward expansion of the barbarians, from 434 CE on, Attila (406–453 CE), the King of the Huns, passed through the Balkans and Greece. Only the walls of Constantinople stopped him in the east. In 451 CE, Attila crossed the Rhine River as the leader of an army of Germans and Huns. There, he was met and stalemated by a mixed army of Romans and Germans, where he was checked for the first time, which saved Gaul from the losses inflicted on the Balkans. The following year, however, he and his men rode into the Po Valley, all the way to Milan. In 452 CE, the Romans sent Pope Leo I (400–461 CE) to meet with Attila, hoping to persuade him to turn back from conquering Italy further. Turn back they did; they crossed the Alps and returned to the Hungarian plains.

The historian, Kyle Harper, speculates that the retreat was the result of "famine and some kind of disease," that is, "a predictable biological consequence of intruders of the steppes colliding with indigenous disease ecology" and not the ability of the Romans to halt his advance. Harper is confirmed in part as tree-ring analyses indicate that the climate of 447, 451 and 452 coincided with extremely dry summers in the Carpathian Basin, which would have decreased the harvests and grazing for animals beyond the alluvial plains of the Danube River and its tributaries. Attila's raids may have been driven, in part, by the need for acquiring food.

The western of the Roman Empire, after the multiple barbarian intrusions and that which followed in Attila's wake, was unrecognizable from the one before Rome's loss at Adrianople. Rome became totally disconnected from its central administrative structures. The last Roman Emperor of the west, the youthful Romulus Augustulus, was deposed in 476 CE by the German warrior Odoacer (432–493) who subsequently ruled Italy proclaiming himself King. In 493 CE, Odoacer was forced to surrender to the monarch of the Ostrogoths, Theodoric the Great (454–526 CE), who continued ruling the Kingdom of Italy until his death.

The year 476 CE is formally viewed as the end of the western Roman Empire. The sequence of events that set the collapse of the most powerful western Empire in motion, lay in the steppes of Asia and the subsequent overwhelming assault of barbarian migrations pushed by the Huns on the westward frontier.

Throughout two pandemics and the assaults from the frontiers, the Catholic Church unexpectedly found itself the most affluent landowner in history and the most powerful. The church and state became Christendom, which had been forming for a century. The church essentially expanded into a vacuum generated by the decay of Imperial Control.

Justinian, Reconquest, Global Cooling and the Plague (541 CE)

By the mid-sixth century, the city of Rome had collapsed. With the city's population decreased from a million in the fourth century to 30,000, the pope took over all administrative functions. In contrast, the centripetal forces that moved the center of the Roman world east continued and Constantinople, buoyed by the wealth of the eastern provinces, prospered. The future looked bright at the east end of the Roman Empire.

Unlike Rome, Constantinople's population numbered half a million people. Fueled by food from Egypt's bread basket, it was the global hub of the known world. In 517 CE, the aged Justin (450–527 CE) ascended through the army and when the eastern Roman Emperor Anastasius (431–518 CE) died, Justin was elected as the successor. Later, his adopted nephew Justinian was groomed for emperor. He assumed the position in 527 CE and ruled until his death in 565 CE. This transition illustrated a relatively stable succession process.

Like Diocletian and Constantine, Justinian, together with his wife Theodora, was an energetic reformer. Primary among his many lasting accomplishments was the landmark codification of 1000 years of Roman Law into a systematic, coherent whole. As a patron of Christianity, he built many churches, poor houses and hospitals, as well as cisterns, aqueducts and granaries throughout the empire. His greatest architectural achievement was completing the redesigned Hagia Sophia 537 CE in Constantinople. Its new vaulting structure lifted a 31 m diameter dome 55 m above the floor. It remained the largest dome in the ancient world and was a great reminder of Roman mathematical and engineering skills.

Militarily, Justinian was motivated to reconquer the lost provinces of the west. In 533 CE, he launched General Belisarius to recover North Africa from the Vandals. This assignment was easily completed and the region remained in Roman hands until the Islamic conquests of the eighth century. Three years later, in an attempt to bring Italy back into the empire, he reconquered Sicily, Naples, Rome and Ravenna, but not the rest of the peninsula. The war with the Ostrogoths was not as easily won and dragged on for a decade with little long-term gain. The result was ruined cities and a rural population reduced to poverty. In contrast, the Eastern Empire still controlled the outposts of Ravenna, Rome and regions of southern Italy for many centuries.

In the eastern provinces, the region of Anatolia with its large population growth, produced a line of cities facing the Aegean to become one of the most urbanized regions of the empire. Likewise, Syria and Palestine were both economically and spiritually vibrant. Cities like Antioch and others facing the Mediterranean Sea south to Alexandria thrived. The region referred to as the "Holy Land" enjoyed a late building boom of monastery and church construction. The Levantine traders supplied ships to ply the waters into the western Mediterranean Sea. Behind the coastal cities, the semi-desert regions from Damascus and Jerusalem to Petra flourished. From Egypt to Carthage, as well as those regions adjacent their fertile hinterlands, prospered. Essentially, all the lands connected to Constantinople advanced after the western part of the Roman Empire collapsed.

This affluence changed after 530 CE. Concurrent with these activities, the Late Antique Little Ice Age arrived in the sixth and seventh centuries CE. The coldest temperatures occurred between 530 and the 680 s CE, with the 536–545 CE years being the chilliest decade in the past 2000 years. That period was due, in part, to global cooling initiated by massive volcanic activity that blocked sunlight. Some of that activity has been identified recently as cataclysmic eruptions in Iceland in 536, 540 and 547 CE. Not only did the cooling have an adverse effect on harvests but it is presumed to have triggered a pandemic in the far east.

This pandemic became known as the Justinian Plague and was due to bacterium yersinia pestis, commonly referred to as a bubonic plague. The rapid cooling may have caused fleas carrying the bacterium prevalent in rodents of southeastern Asia to migrate to the native black rats of the region. From then on, the trading networks linking Asia along the Indian Ocean west to the Red Sea carried the rats north through the Red Sea to the trading networks of the Roman world. The story was similar to the Antonine Plague. The black rats were prolific breeders who loved grain and were easily transported through these corridors.

In 541 CE, the Justinian plague first appeared in Pelusium, which lay on the Mediterranean Sea due north of the northern Red Sea port of Clysma. There, the contagion divided into two different directions. One traveled west to Alexandria, where it spread to the rest of Egypt. The other traveled east through Palestine, where it infected Jerusalem and continued north to Anatolia.

The plague was diffused at two speeds, one migrating slowly along Roman roads, the other spreading rapidly along the shipping lanes of the Mediterranean Sea. In February 542 CE, the plague reached Constantinople and lasted four months. By 543 CE, the plague reached Carthage, Rome and Arles and finally reached Britain in 544 CE and Ireland in 546 CE. The Roman networks of communication and trade was the perfect conduit.

Although the urban communities were hardest hit, the pandemic also spread throughout rural villages and the countryside. Grain was left unharvested and grapes and olives were left to rot. The climatic vicissitudes prior to the plague had reduced the supply of food and weakened the peoples' immunity, which allowed the plague to easily attack the hungry weak citizens. Approximately half the population affected by the plague died.

Prior to this time, the Justinian Plague was the greatest mortality occurrence in human history. The effect of the plague crippled Justinian's goal of reunifying the former empire, Which fractured in pieces. The human loss was similar to that experienced later, during the fourteenth century, where 40–60% of all the people in Europe, North Africa and the Middle East would die from the Black Death.

Over the next two centuries, the plague reappeared seven more times in Constantinople but never returned after 747 CE. It left as mysteriously as it arrived in 542 CE. The subsequent reoccurrences did not come from outside the empire but rather from the seeds left behind from the earlier disasters. The last visitation was wider in geographical extent than the first, with the losses forcing the emperor of Constantinople to repopulate the city by involuntary immigration.

Eastern Roman Empire Transforms to Byzantine

The wetter and colder climate changes and series of plagues after 541 CE, reversed the gains made and limited the building boom to churches, which were motivated by the fear of the coming end of the world. The demographic decline and simultaneous loss of resources, from a decimated tax base, significantly reduced the ability of the imperial empire to pay and field the army, required by its extended geography.

The emperors after Justinian had decreased ability to prevent encroachments into its territory with each reappearance of the pandemic and the pointless war with the Persians. The Roman state was continually reduced and weakened. The Persians had hoped to reconstruct the earlier Persian empire of the Achaemenids. In pursuit of this objective, the Persians acquired Mesopotamia from the Romans in 607–610 CE. Syria was lost in 611–613 CE, Palestine in 614 CE and Egypt fell in 616 CE. In 628 CE, however, the losses were briefly reversed under Emperor Heraclius (575–641 CE).

In Arabia, the rise of a new monotheistic religion that bound together groups of believers, mending tribal divisions, called Islam, was developing. It was under the leadership of a new prophet, Mohammed (571–632 CE), who proclaimed that there is but one God, Allah. In 634 CE, Mohammed's followers traveled north from Medina and in 3 years destroyed the forces of the Romans and Persians. Unable to hold off the Arabs, in 636 CE Heraclitus retreated to Anatolia, leaving all the territories he had reclaimed earlier from the Persians. These territories included the cultural and spiritual jewels of the empire. In 637 CE, the Persians retreated, to the Iranian plateau. Subsequently, in the next two decades the Arabs pushed both west along the southern frontier of the Mediterranean Sea and east to the eastern border of Iran, absorbing the Persians and forming an empire to match that of the Romans, called the Arab Caliphate.

During Heraclitus reign (610–641 CE), the Empire's administration and military were restructured with Greek replacing Latin for official use. The character of the Empire become more Greek than Latin. Underlying Constantinople was the Greek City Byzantium. To mark the transition in character, historians introduced the term Byzantine for the reduced Roman Empire even though the peoples of the region continued to call themselves Romans.

By 650 CE, the Roman Empire was reduced to a small amount of territory north of Constantinople, Anatolia, Carthage with its associated territories and pieces of Italy. The prized possessions of the remaining Roman Empire were conquered by the Arab invaders. These rapid losses illustrate the damage the climate, the plagues and the endless wars the Persians inflicted. The fragmented Roman Empire never reunited. Constantinople, however, remained a pillar of Christian faith and resisted the later Ottoman Turks until its walls were breached in 1453 CE. This final loss of the eastern provinces would later represent a major change in the course of history.

Chapter 7
Europe's Beginnings

The administration of the Romans represented an amalgamation of many polities with their own culture, language and history, including: the Greeks, Phoenicians, Egyptians and Celts. This integration, which formed the beginnings of western Civilization did not last in this form. It started with the progression of cultures in Mesopotamia and Egypt, extended to Greece and Rome and expanded throughout the Mediterranean basin to Europe. With the ending of the western Empire (476 CE) and the contraction of the eastern Empire (650 CE), the mantle of western civilization passed to Christendom in Europe and south to the emergence of Islam. It is to the first, which eventually led to western science that is now followed.

Historians have divided this early European period into three parts: Early Middle Ages (500–1000), High Middle Ages (1000–1300) and Late Middle Ages (1300–1500).

Early Middle Ages: 500–1000 (Kingdoms)

Barbarian Kingdoms

In 476 on, Germanic tribes continued advancing into Roman territory. By the sixth century the former Roman territories became divided into numerous barbarian kingdoms. The Visigoths commanded much of Gaul and Spain and a kingdom of Franks controlled a region in the north of Gaul. Smaller Burgundian and Alemanni states administered the center of western Europe and Saxons and Anglos settled the far northwest adjacent and distant from Britain. Brittany was managed by refugees from Britain together with the former provincials and Italy remained in the hands of the Ostrogoths. At this juncture, there was little stability among the emerging kingdoms and conflicts were frequent.

© The Author(s), under exclusive license to Springer Nature Singapore Pte Ltd. 2024   87
T. Sanford, *A Whirlwind History of the Universe and Mankind*,
https://doi.org/10.1007/978-981-97-2674-5_7

At the time of these barbarian occupations, most Europeans had considerable exposure to Roman culture and were Christianized. The new kingdom courts became like those of the provisional governors, where kings behaved more like Roman magistrates, not emperors. Before the final breakdown of imperial Roman authority, this behavior eased friction between these new kingdoms as they remained part of the Roman Empire.

The strongest warriors of the new kingdoms formed the ruling class. Persuading many aristocratic families to accept the evolving leaders helped preserve their wealth and smoothed the transition for both. The kingdoms created law codes that maintained a well-defined difference between the barbarian settlers and the original population. The law codes were basically Roman but with provisos that institutionalized the superior status of the occupiers over the provincials. In some cases, as in Italy under Theodoric, the Roman tax system and government were left untouched.

Assimilation of the barbarians was slow. In the long term, the language and culture of the indigenous populations proved the most lasting. Latin evolved to Italian in Italy. The Visigoths of Spain eventually become Latin speakers who transitioned to Spanish with clear Latin roots. The Franks of Gaul similarly developed into speaking French with well-defined Latin origins. The Anglo-Saxons of Britain were an exception; there a Germanic language remained spoken. Administrative writing and literature, however, continued in Latin. Additionally, the Catholic Church maintained contact between all the kingdoms and fostered the use of Latin everywhere.

The new rulers respected the church. It assisted keeping the provincials at peace with their rulers. Bishops often would be the only Roman administrators having the capability of providing guidance that could maintain law and social peace for the barbarian rulers. The kings held the bishops in high regard, especially the Bishop of Rome. Over time, the Roman Bishops assumed the role of a former western Roman Emperor and the titles and ceremonial behavior of the emperors were adopted. In essence, the Catholic Church became an exact replica of the imperial Roman bureaucracy.

Although aspects of the previous culture survived, the new kingdoms were fiercely independent. Occasionally they traded with one another and other times they fought. They all had diplomatic contact with Constantinople but were not subject to its authority. The disintegration of the western Empire did not pose a concern to the eastern Empire; rather, they were focused on interacting and courting the individual kings.

Similar to the eastern Empire, the effect of the plague, climate and war had a large effect on the former western Empire. The population declined but never disappeared and life was greatly simplified. In Italy, many towns shrunk or disappeared altogether. The manpower reduction limited the recovery from nature's destruction as ports silted in and terraces washed out. Estimates suggested the population of Italy was reduced to half or a fourth of the Roman numbers. Coins disappeared and were replaced by bartering. Simple household items from abroad vanished and the wealth of the aristocracy evaporated. The middle class could no longer reconstitute itself and the Church inherited a less prosperous world. Settlements on the fertile lowlands retreated to hilltop villages. The countryside reverted to conditions prior to the early founding of Rome itself. From Italy to Britain, the situation was very similar.

By 100 years after the fall of the western Empire, the territory of modern day France, Germany and the Netherlands evolved into the realm of the Franks. Southern Britain came under the control of the Anglo Saxons and Spain evolved into the kingdom of the Visigoths. The Ostrogoths of northern Italy became the kingdom of the Lombards. They were a Germanic tribe who had invaded Italy in 568 from the north. Other parts of Italy, including Sicily, Sardinia and Corsica, returned to the eastern Roman Empire.

The Franks

In 465, the Frankish King, Clovis (465–511), was born in Tournai, Belgium and died in Paris. He inherited the kingship from a subgroup of Franks in 481. By the time of his death, he had united and extended Frankish rule south into southern Gaul, pushing the Visigoths back into Spain and established the Merovingian Dynasty. By 536, the Merovingians had succeeded in uniting the territory of modern day France into Frankish control.

Clovis' previous conversion to Catholicism helped other Goths to follow his transformation. Many of the barbarians had been introduced earlier to an alternative form of Christianity called Arian by evangelizing missionaries who espoused a different view of the Trinity. Throughout Gaul the Franks adopted Latin, encouraged an urban lifestyle and maintained a Roman form of government. With time, the parts of Gaul settled by the Franks became known as Francia.

Within Merovingian Francia, the Merovingian kings frequently divided their territory among their sons upon passing. The eventual result was a territory divided into a number of different kingdoms (such as Neustria, Austrasia, Burgundy), which often warred with one another, thus weakening the dynasty itself.

Internal to Austrasia was an up-and-coming Carolingian dynasty. Inside the dynasty was Charles Martel (688–741) who distinguished himself in a Frankish civil war that lasted from 714 to 719. Earlier, between 711 and 716 the Arabs and Berbers of north Africa had conquered most of Visigothic Spain and for a brief period conquered the regions of southern Francia. Charles Martel, at the Battle of Poitiers in 732, however, thwarted the Arab advance north of the Pyrenees, saving western Europe from the Arabs. He continued military expansion and by 739, the territories of southern and western present-day France, fell under Carolingian rule.

In 747, Charlemagne (747–814) of the Carolingian dynasty was born in Aachen, Germany. He became the sole king of the Franks, waging war endlessly during the first decades of his reign in 768. Early in his kingship, Charlemagne initiated a long war on the pagan Saxons living to the east of the Rhine River, extending the Frankish frontier to the Elbe River. This territory had never been conquered by the Romans.

From 773 to 774, Charlemagne conquered Lombard, Italy taking the title King of the Lombards. Henceforth, northern Italy became ruled by the Carolingians, with central Italy remaining under the control of the Popes and southern Italy under the Byzantine. By such actions, Charlemagne expanded the Frankish Empire to its greatest extent.

On Christmas Day in 800, Charlemagne was crowned Emperor of the Romans in the Old St Peter's Basilica in Rome by Pope Leo III. The coronation was a momentous event as no leader had united the bulk of western Europe, since the fall of the western Roman Empire. Because of this unification, he has been called the father of Europe.

During the reign of the Carolingians, the secular authorities controlled the religious authorities. They relied on the clerics to fill the governmental bureaucracy. Appointing them to positions within the church was a method of fostering loyalty. The Carolingians extended the parish system through Europe, in an attempt to ensure that all Christians had contact with priests and could hear religious services. By creating a mandatory tithe, a tenth of each families personal income to their local church, the church was put on a sound financial footing. The Carolingians supported Christian missionaries, in particular those proselytizing to the German pagans as a necessary means to gain support of the Carolingian authority.

Of importance to Europe, Charlemagne was motivated to learn and as a role model attracted scholars from all over Europe to his court at Aachen. This assemblage initiated the Carolingian Renaissance and accelerated the development of intellectual energy within the western Church. The number of schools increased throughout the empire. One focus was to restore and establish uniform renditions of important religious texts, like the Bible. Another was to raise the basic level of education, especially that of the clergy.

Old copies of these religious texts were found, collated with similar copies, which removed errors and were recopied. In the process, classical pagan literature was used principally to better understand the meaning of Latin. Throughout this process, many works of the classical authors, like Cicero or Virgil, were saved from destruction. In these efforts, the Carolingians developed a uniform and well-defined hand writing referred to as Carolingian Minuscule, which is similar to that used today. In the past, there were no intervals between words or sentences or rules for punctuation. As a result, a clear distinction between Latin and the vernacular languages of Italian, French and Spanish developed by the ninth century, with the vernacular being recognized as official languages.

This time period also saw the transition from the large Roman estates to the medieval manor. The Roman estate consisted of two parts: the demesne and the tenancies. The demesne was land that the estate owner directly farmed, using wage and slave labor. The tenancies were land rented out. In general, the demesne was substantially larger than the tenancies. On the medieval manor the demesne was farmed by serfs, not slave labor. Along with the development of the medieval manor house and serfdom, the Carolingians introduced feudalism throughout Europe.

Predecessors of France and Germany

As in the collapse of the Merovingian dynasty, competition among the inheritors of the pieces of the Carolingian dynasty in the ninth century resulted in the devolution of its authority. Under Charlemagne, members of the aristocracy often served as local officials known as Counts and Dukes. After 838, counts ignored Carolingian authority and made themselves free of control within their counties. Many took powers reserved for kings and made their offices hereditary. In 843, under the treaty of Verdun, the west–east division of the Frankish Empire gradually developed into the creation of separate kingdoms, with west Francia becoming the predecessor of France and east Francia the predecessor of Germany.

In the east, Carolingian control was lost in 911 to the Ottonian dynasty. There, Otto I (912–973) formed a strong German State out of east Francia and the German speaking lands by controlling of his wayward vessels and his victory over the aggressive Magyars (Hungarians) at the Battle of Lech in 955. Otto I expanded his authority over northern Italy and later the kingdom of Burgundy, becoming the most powerful European ruler after Charlemagne.

In 962, Otto I had himself crowned Emperor in Rome by Pope John XII. His crowning signified the birth of the Holy Roman Empire. Like the Carolingians, the Ottonian rulers continued to control the church and the papacy. And like Charlemagne, Otto I supported education, which helped keep learning alive during medieval Europe, resulting in the Ottonian Renaissance.

In the west, the Capetian dynasty replaced the fragmented French Kingdom's Carolingian dynasty in 987. The Capetian dynasty remained in control from 987 to 1328, which permitted the kings to slowly build on the successes of their predecessors. To facilitate this, the kings established a practice of securing the election of the next king while still being alive. Additionally, the Capetians practiced primogeniture, which further enabled heirs to inherit the majority of a dynasty's assets. Eventually, Philip II Augustus (1165–1223), was able to restore the lands lost to England through marriage and inheritance and thus French royal power over western France was restored.

Britain Becomes England

In 43 under the reign of Clausius, the Romans invaded Britain, which at the time were collections of illiterate Celtic tribes. In 410 the Romans left, as the western Empire was beginning to crumble and the Roman forces were needed elsewhere.

After the Roman departure, Anglos, Saxons and Danes began first to raid Britain and then to settle in specific areas. By 600, the eastern and southern regions of Britain came under control of the Anglo-Saxons. The areas to the west and north, like Wales, Ireland and Scotland, however, evaded their advancement. In the process, the Anglo-Saxons formed a number of kingdoms along the eastern and southern coast of Britain.

From north to south, they became the Kingdom of Northumbria, the Kingdom of Mercia and the Kingdoms of Wessex and Sussex.

During this period, some of the indigenous Celts left and settled in present day Brittany across the English Channel. As these movements were occurring, the Scotti from Ireland migrated to the British west coast and later went on to conquer present day Scotland. Throughout this time, Roman villas and towns were abandoned. The economy reverted back to a pre-Roman state of barter and illiteracy. The Germanic language of the Anglo-Saxons became the language of Britain now called England (land of the Anglos). Christianity disappeared with the indigenous peoples adopting the Anglo-Saxon paganism. In the Celtic borderlands Christianity remained.

In the seventh century, Papal and Irish missionaries reintroduced Christianity to Anglo-Saxon England. By the 660 s, Christianity was generally accepted by the Anglo-Saxon kings and people. As the bishops and their associated retinues returned, small-scale urbanization followed together with literacy. Now in reverse, Anglo-Saxon and Irish monks spread out across to the continent to the Franks, bringing religious reforms and establishing monasteries. Centers of intellectual learning were formed. The message of the monks was well received by the Frankish aristocracy who gave land and money to the monasteries.

During the eighth and ninth centuries, the Viking (Danes) raids on England became more disruptive. In 865, the Danes conquered the lands from Northumbria south to the Kingdom of Wessex. The Kingdom of Wessex survived under the leadership of its king, Alfred the Great (849–899), who was born at Wantage, Berkshire. He learned Latin and promoted literacy and education. He translated Latin books into English and promoted codes of law with special consideration for the weak and dependent. He fostered a program of education through the intellectuals attracted to his court, similar too Charlemagne. In 886, King Alfred reoccupied London, which had been abandoned by the Romans in the fifth century. By 959, the rest of the Danish kingdoms were conquered by later kings of Wessex, thus forming a united Anglo-Saxon England.

Anglo-Saxon England did not last long. The Danish absorbed England in the early eleventh century, forming part of a Scandinavian Empire. In 1066, under William the Conqueror (1028–1087), the Normans defeated the Anglo-Saxon English at the battle of Hastings, with the full conquest of England taking another five years. The Normans were descendants of the Vikings who arrived earlier and settled in the region of Normandy, during the beginning of the tenth century. They had been living in France so long that their language had evolved into French.

Prior to this time, English culture was influenced by the regions of the North Sea with links to Denmark and Norway. Under Norman control it was ruled by French speaking aristocracies for the next centuries, which were orientated to Normandy and Europe. This transition changed England completely. The Anglo-Saxon nobility was replaced by the Normans and a feudal system was imposed. The lands taken by William the Conqueror were redistributed to his supporters. William reorganized the church, bringing in clergy from France to be bishops and abbots. Castles (i.e. the Tower of London), cathedrals and monasteries were erected. Because of previous land ownership issues between the French and English nobility, the history of the two countries became deeply entwined. The English language evolved to be a mix of both Anglo-Saxon and Norman-French words.

Spain Becomes Al-Andalus

In contrast, Visigothic Spain with its center at Toledo, maintained a significant cultural identity with its Roman past until its collapse, between 711 and 716, by the invading Arabs and Berbers. In 756, Abd al-Rahman I, an Arab of the Umayyad dynasty centered in Damascus, fled to Spain and completed the Arab expansion to the far southwest. He initiated the political independence of al-Andalus (as Spain became known) and established the Emirate of Cordoba. Successors established a Caliphate of Cordoba in 929 and the Umayyad Dynasty ruled until 1031. During the conquest, the Christian Visigoths were pushed northward and by 1000, they formed the Kingdom of Leon, Kingdom of Navarre and County of Barcelona in the north of Spain.

The Arabs rejuvenated Spain, bringing in new agricultural practices, plants and technologies. The experience of the Arabs, in their arid environment, was applied similarly to dry al-Andalus. The Arabs introduced new technologies in water extraction and management, such as: waterwheels, local water-raising machines, quants (underground tunnels) and dams. Along with the increased use of water extraction and crop rotation, new plants were introduced. These included rice, citrus fruits, cotton and sugarcane.

For centuries, Al-Andalus was home to large numbers of Muslims, Jews and Christians, who lived in harmony. Although Jews and Christians had to pay a special tax, they could practice their religion. Of importance to Western Europe, al-Andalus became culturally connected to the trade and commercial activities of the Arabic (Muslim) world that stretched across the entire Middle East, indirectly coming into contact with India and China. Spain had a marketable source of metals, food and wood that was exchanged for the luxury goods of the east.

Al-Andalus was more urbanized than the rest of Western Europe. At its economic and cultural center, Cordoba was larger than any city in Western Europe, with a population of 100,000 in 1000. The city became a center of learning. The Umayyad rulers collected numerous Greek and Islamic manuscripts, building up great libraries. Exemplary was the polymath Ibn Rushd (Averroes) (1126–1198), who was born in Cordoba and wrote over 100 books on philosophy, medicine, astronomy and physics. He was a great promoter of Aristotelianism, defending the rational as the basis of wisdom and knowledge over religious beliefs. Translations of the works of Averroes into Latin reawakened western Europeans to Aristotle and other Greek thinkers, who were forgotten after the fall of the Roman Empire. Owing to the many Muslim kingdoms of al-Andalus vying for power, the Caliphate ended in civil war after 1031. And Cordoba eventually fell to the Castilian King Ferdinand III, becoming part of Christian Spain in 1236.

High Middle Ages: 1000–1300 (Recovery from the Dark Ages)

Improvements

Between 1000 and 1300, the population of Europe increased, doubling to 50–100 million from a nadir in the seventh century. Life expectancies increased 40%, from 25 to 35 years. This growth, relative to the previous years, was due to the absence of several negative processes prior to 1000 and to a number of positive forces enhancing expansion after 1000.

Reducing the population was the Bubonic Plague, which periodically reoccurred after its first appearance in the mid-sixth century and ended during the mid-eighth century. The Arab expansion into the south, Viking invasions in the west and Hungarian raids in the east, during the ninth and tenth centuries, displaced a significant segment of the European population.

On the positive side, these incursions ended by 1000. The introduction of the heavy metal plow enabled the efficient turn-over of the deep northern European soil, increasing agricultural production. The soil was now plowed, using horses with collars rather than oxen. The collar pressed against the shoulders of the horse, not choking it, and allowed the horse to apply 50% more power to a task than an ox could. Nailed horseshoes, along with better breeds and improved feed, increased the endurance of the horse. By 1200, these improvements significantly increased the development of transportation and trade. This reduced the cost of conveying heavy goods on land, relative to the Roman era, by a factor of three. The watermill, with the multitude of rivers available, began to be used on a large scale, again increasing food production. In addition, the relatively dry warm conditions between 950 and 1250, known as the Medieval Warm Period, further aided productive agriculture.

Prior to 1000, European society was still feudal and divided into three classes: a warrior aristocracy, a clergy and field laborers. The increase in food, however, encouraged greater labor specialization. Farmers and others developed new skills, resulting in the emergence of a merchant class within towns. As towns grew, new opportunities developed and a fourth class evolved, the townspeople.

Towns

In parts of the Mediterranean region, particularly Italy, southern France and Spain, urban life continued. There, the nobility often lived in towns for a fraction of the year. In Northern Europe, the nobles usually lived in the countryside not in the towns. Within a town, Jews were the largest non-Christian minority. A protective wall often separated a Jewish quarter from the rest, especially in Mediterranean locales.

Outside these regions, urban living in Europe was at a low point at the turn of the millennium. Europe was economically underdeveloped, literacy embryonic and geopolitically a backwater. Northern Italy contained the largest European towns second to Spain, with typically 10,000–20,000 people. In northern Europe, the largest towns contained only 4000–5000 people. Owing to the demographic improvements after 1000, northern Italy developed cities of 100,000–200,000 people and by 1300, the population of northern European cities expanded to numbers of 40,000–50,000 people.

This renewal of urban living had enormous consequences. Foremost was the increasing sophistication of commercial life and the increased trading contacts between Europe and the outside world. At the forefront of that development were the Italian merchants, transporting goods between the eastern Mediterranean Sea and the rest of Europe. These goods were sold at fairs within Europe and motivated the revival of a monetary system that was once part of the Roman Empire. Arab numerals replaced Roman numerals, and introduced the number zero, all of which greatly increased the efficiency of the counting system. Because of the need to use money, sign contracts and keep records and accounts, the townspeople revived lay numeracy and literacy. With the growth of urban life, diversity in occupations evolved and flourished, giving rise to a middle class.

During this early period, urban society was stratified economically and strictly regulated. In Italy, for example, merchants and merchants who had grown rich enough to live off the land they had purchased were referred to as populo grosso. People who worked with their hands (retail merchants, artisans or farmers) were referred to as populo minuto. Town governments were generally managed by the populo grosso.

Because of the inability of the royal governments to quell noble violence within a town, the towns established communes for mutual defense to maintain peace. Commune officials were created to oversee the commune. They were elected by the townspeople and typically served for a year. In Italy, these officials were known as consuls. Those who injured a commune member were judged by the consuls. If those guilty refused to accept the council's judgment, subsequently vengeance was exercised. In northern Italy and northeastern France, the communal movement grew, during the late eleventh and early twelfth centuries and eventually spread to the rest of Europe. In England, where the kingdom was powerful, communes were unnecessary and never formed.

Just as the communes protected the people of a town, guilds protected the members of a given trade from competition and economic hardship. Often as many as 100 different guilds existed within a town, one for each trade: baker, carpenter, shoemaker, tailor, etc. Membership was required to practice any trade. The guilds standardized the making and selling of their products. The guilds often determined the tools, techniques, materials and working hours. Quality standards were set and enforced. Guilds provided a social net to help a member and his family if the member become ill or was injured or died.

For peasants working on manors, the towns provided an escape. Serfs who managed to migrate into a town and met the citizenship requirement (generally, living there for a year and a day) became free. The fear of peasant loss to the towns motivated the nobility to reduce the burdens of lordship over time. By 1300, most peasants were free, although in some parts of Europe, serfdom continued to exist into the modern period.

Clocks

During the latter half of the thirteenth century, the growth of towns with their urban mercantile populations generated a demand for improved timekeeping devices. Likewise, precise observance of prayer times by monastic orders motivated the requirement for reliable time announcements. In 1283, one of the early accurate weight-driven mechanical clocks was installed in Bedfordshire, England at the Dunstable Priory. They were typically powered by the gravitational pull of heavy weights falling slowly and often attached to a mechanism striking a bell. By the end of the thirteenth century, craftsmen were generating clocks for cathedrals and churches throughout Europe. The name derives from the Latin word for bell, *clocca*.

Woolen Cloth

In Flanders, weavers using looms began producing reasonably priced woolen cloth of superior quality compared to the common homespun. This new cloth gained a reputation, initially locally and later as far away as Italy. The demand for this cloth increased to the point that it could not be met by the local sheep farmers.

England provided the answer with its sheep that produced wool of high quality. There, farmlands were reorganized to meet the Flanders need. By 1200, England became the primary supplier and Flanders the manufacturing center of a lucrative industry. Cloth made in Flanders was selling in northern Europe as well as in Italy. Towns manufacturing cloth, like Ghent, and those that stored and distributed the finished goods, like Bruges, boomed.

The woolen cloth industry had a beneficial effect in Europe. Not only did regions surrounding England and Flanders grow but port cities like Genoa and Venice did as well. The merchants discovered that these northern woolens were also sought after in the eastern Mediterranean region. Transport to the east further increased the profits of the merchants. In 1000, the port and the surrounding interior Italian cities in northern Italy had fewer than 5000 people. By 1200, however, the cities experienced an economic growth, as big as that in Flanders. Venice and Milan, in particular, expanded to populations in excess of 50,000, becoming the largest cities in Europe.

The Flemish textile industry flourished until the Italians entered the industry as producers. Initially the Italian cloth was of poor quality. But by 1220, they produced cloth that was fine enough for export and, 100 years later, they were generating fabrics of similar quality to those of Flanders and the Levant. By 1320, the Flemish textile production declined as the industry moved to Italy.

Gothic Cathedrals

European prosperity and peace in the early twelfth century allowed a Gothic style of architecture to evolve out of the Romanesque style for churches and cathedrals. The Gothic style represented a major distancing from the basic construction that prevailed. This style brought more light into interiors. The dramatic heights achieved and new light that radiated in from the heavens reinforced the power of the church.

A new feature of the Gothic style was the pointed arch, which was likely borrowed from Islamic architecture. The pointed arch relieved some of the lateral thrust on the walls, which reduced the stresses throughout. That change allowed the size of the interior columns to be scaled back and ribbed vaulting lightened the ceiling, reducing more stresses. Flying buttresses added structure to support thinner walls, which enabled greater heights. The results were elegant structures with soaring interiors, with large stain-glass windows. One of the first cathedrals to use flying buttresses was the Notre Dame Cathedral in Paris, for which construction was started in 1163 and completed in 1345. The building of these refined cathedrals was a testament to the evolving independence of Europe from the past and the enhancement of its new technological and mathematical skills.

Scholasticism

For most medieval people education was not deemed necessary. The study of the Bible and the writings of the church fathers formed the center of most monastic teachings and early European education. Learning and understanding Latin was emphasized. The monastic teaching was passive in nature in that the students (monks) were not to question the texts being studied, just as they were not to question the authority of the abbots.

By the end of the eleventh century, this meditative and associative form of teaching evolved into the scholastic method, which, in contrast to being passive, was based on argumentation. This new teaching method flourished predominantly in cities. Urbanization fostered centers of communication with discussion, debating and bargaining, during the course of a normal workday. Scholasticism in the classroom paralleled the new urbanization. It examined problems on which experts differed and attempted to resolve through philological analysis and formal logic based on revered texts like the Bible or those of the church fathers. Scholasticism was not a set of beliefs but rather a learning technique.

Although these techniques differed substantially from the earlier monastic methods of learning, all were based on the study of ancient texts, which assumed the correctness of the old authorities and were not questioned. Hence, scholasticism did not really advance human knowledge. By 1300, however, the use of logic to resolve an argument or discrepancies in revered texts was liberating.

The church welcomed the development of scholasticism, since it provided better techniques to fight against heresy. With regard to non-Christian authors, the monks felt that their study would improve the mind, the ability to read Latin, and be beneficial in reading and understanding the Bible. Classical authors like Cicero, Plato and Aristotle and the works of the Muslim author Averroes were deeply appreciated. Aristotle, along with other classical authors, was not readily reconciled with the church doctrine on a number of points. He, for instance, believed that the world had always existed and would continue to exist and that there was no immortality of the soul. Accordingly, his works were often forbidden to be taught by the ecclesiastical authorities, however, these restrictions were rarely followed.

Universities

In ancient Greece, higher learning existed long before universities were beginning to be established in medieval Europe. The Platonic Academy was founded in 387 BCE and Aristotle's Lyceum was founded in 335 BCE. These were not institutions of mass learning. The Great Library of Alexandria, which was part of the Museum in Alexandria, Egypt functioned as an international repository of knowledge from the Hellenistic world and beyond. It operated like an international university, where students were taught by the best educators.

During the Romanesque period, scholarship followed Platonic traditions established by the Greeks. Sons of aristocratic Romans were educated locally by Greek teachers or sent away to be educated by Greeks in Athens or Alexandria. By the sixth century, these institutions were shut down, owing to their connections to pagan philosophies of the past. With the end of the Roman Empire, higher scholarship transitioned to isolated monasteries, where monks and priests initiated the documentation of past knowledge from the Greeks and Romans.

With the growth of towns, monasteries opened cathedral schools operated by bishops. The Sorbonne, for example, evolved from an early monastic cathedral school in Paris. Bologna offered the first true university, which opened in 1088. Others soon followed and, by 1300, approximately 20 universities were operating. These were initially corporations (scholastic guilds) of students and teachers without fixed buildings. In contrast to wandering scholars opening their own schools, the church authorities advocated founding universities, where controlling the teachers was easier.

The curriculum was based on early Greek and Roman structures of education. Four faculties defined the European university: that of the arts, theology, law and medicine. In the faculty of arts, the trivium and quadrivium were studied. The trivium dealt with grammar, rhetoric and logic. The quadrivium consisted of arithmetic, geometry, astronomy and music theory. These formed the foundation of all education prior to the study of theology, law or medicine.

A degree in theology was considered to be the most prestigious. Often, sources of individual books were offered, like *Physics* by Aristotle. Six years were commonly required for a bachelor degree and twelve years for a doctorate. All courses were taught in Latin. The current university system of degrees has remained essentially unchanged: bachelor of arts, master of arts and doctor of theology or law or medicine. In contrast to today, the university was only open to men.

Crusades

By 1000, a large fraction of the Byzantine empire had been lost to the Arabs but the empire still retained Anatolia. Between then and the mid-eleventh century frequent upheavals had occurred in the Muslim world. Most notably for Byzantium, the Turks (now converted to Islam) migrated from the plains of Central Asia to the Middle East, forming the Seljuk Sultanate. The Seljuks were the first of the nomadic empires to dominate the region. In the Battle of Manzikert, fought in eastern Turkey circa, 1071, the Imperial army of the Byzantines lost decisively to the Seljuk Turks. The result was the loss of Anatolia to the advancing Seljuks.

In 1095, the Byzantine Emperor Alexius I, appealed to the Pope for aid in further containing the Turks. Earlier in 1054, a Great Schism had occurred between the two Christian centers (the Church of Rome and the Church of Constantinople). Theological and political differences had developed over centuries. These differences created animosity and led to a clear distinction between the Catholic and Eastern Orthodox Churches. In an attempt to heal the 1054 break, Pope Urban II (1035–1099) was motivated to give aid and help take back the city of Jerusalem, which had fallen under Seljuk control in 1065. Prior to that time, pilgrims had been permitted to travel to the Holy City but now were forbidden to do so.

With these events as background, Pope Urban II promoted the First Crusade in 1095, calling for the return of the Holy City of Jerusalem to the west. The Christians had strong incentives arm them selves and to recapture the city. Urban's speech promised the remission of sins to those who took part. Moreover, if someone died along the way, they too, would have their sins remitted. In addition, earthly rewards would be granted that included forgiveness of debts, freedom from taxes, plunder from conquests, as well as, new political power and fame. From the pope's perspective, the Crusade would help direct the violence of aristocrats in Europe's south to the Levant.

The first Crusade resulted in the recapture of Jerusalem in June 1099. Subsequently, the Muslims unified against the invaders and began a series of religious and political wars; seven more Crusades from the west, continued the conflict. These wars led to the development of Crusader States in Syria and Palestine. In 1291, the Muslims regained control of Jerusalem and the Levant, which remained under Muslim control until the twentieth century.

For Europe, the Crusades opened the Middle East to the west, since the fall of the western Roman Empire. The Crusades created a constant need for supplies and transportation. Ship building and the manufacturing of supplies increased. Seaport cities like Genoa, Pisa and Venice, as well as associated inland communities, prospered as a result of the increased trade.

During the Crusades, new goods and ideas were exchanged. The Europeans were introduced to advanced mathematics, the number system and to medical principles that had been long forgotten. The compass, which was invented in China, made its way to Europe and improved navigation, along with spices, coffee, perfume, soap, glass mirrors and a musical instrument, which became the ancestor of the modern guitar. By the fourteenth century, the Crusades had influenced many aspects of daily life in every part of Europe, from the church and religious thought to economics and politics. The Crusades demonstrated that Europe was no longer a backwater but was able to initiate expansion movements at the expense of its neighbors.

Investiture and Kingdom of Germany

According to cannon law, bishops and abbots were elected by local clergy or monks. In the Holy Roman Empire, however, emperors and secular authorities made the appointments. Church authorities were frustrated by the fact that emperors controlled the papal elections, which generally promoted officials who were only well disposed to the Holy Roman Empire.

Pope Gregory VII (1015–1085), a leading proponent of the movement to bring control of clerical elections back to the church, was driven by the belief that the church was founded by God. The investiture of the eleventh century (with the vassal taking an oath of fealty to an overlord) unfolded when the members of the Gregorian Reform movement wanted to end secular control of the church. In particular, they wanted the ceremony of investiture to end, since it was the most visible symbol of secular control. As a start to that disengagement, the papal elections were held by a college of cardinals in 1059. In 1075, Gregory VII disallowed the enactment of investiture everywhere in Europe. This edict precipitated 50 years of civil war in the Holy Roman Empire. During the process of disengagement, the German emperor declared that the pope should be deposed and the pope, in turn, excommunicated and deposed the emperor. Many of the German nobility rebelled against the emperor. Their motives were not religious. They were interested in constructing stone castles to expand their lordship over the surrounding peasants, without the needed permission of the emperor. In the process, the German nobility forced the peasants to obey their laws and extorted revenue from their servitude.

Eventually, the investiture controversy ended in a truce between the empire and the papacy at the Concordat of Worms in 1122. Under the terms of that truce, the emperors agreed to abandon the use of investiture and permitted free canonical elections of the bishops and abbots. In exchange, the papacy consented to permit imperial representatives to attend and participate in the election of bishops and abbots. This struggle resulted in damage to the Holy Roman Empire. In the Kingdom of Germany and northern Italy, imperial authority never fully recovered. The independent stone castles and associated lordships that had been created were not removed. In northern Italy, the towns moved resolutely in the direction of city-states.

During the next 100 years, the conflict was never fully resolved. Between 1254 and 1273, referred to as the Great Interregnum, the Holy Roman Empire had no emperor. It eventually devolved into a fragile confederation of independent polities: duchies, city-states and episcopal sees. The Holy Roman Empire would persist in this state until 1806, when the emperor stepped down. The Italian communities, the Papal State and now the Kingdom of Sicily (which had come under German control) became independent, evolving on their own.

Magna Carta and Parliament

Prior to the Norman Conquest of England in 1066, England's government was relatively well managed. England was divided into shires and the royal officials were efficient, loyal administrators. The Anglo-Saxon Kings used documents to govern instead of relying on oral tradition. The kings established permanent seats of government. The royal treasury, located at Winchester, was one example. After the conquest, the Normans continued the solid level of administration, strengthening royal authority.

In 1086, William the Conqueror ordered that a national census be compiled in the *Domesday Book* as the first survey of the resources of the kingdom since Roman times. Later, William's son, King Henry I (1068–1135), created the exchequer, a government accounting office where every representative of a shire was required to submit the revenues owed the king. England's next king, King Henry II (1133–1189) was born in Le Mans, France, and as Duke of Normandy (1150), Count of Anjou (1151), Duke of Aquitaine (1152) and King of England (1154–1189), he expanded the Anglo-French domains. He had the instruments of finance, justice and administration revised by his ministers, based in part on Roman Law. All of this assisted in the creation of the *Magna Carta* and became the basis of the future English common law.

During the reign of his son, King John Lackland (1166–1216), most of the French land acquired by England through inheritance and marriage was lost to France. In the process of trying to regain the lands, high taxes were imposed. Conflict with the pope developed, resulting in King John Lackland's excommunication. England was subsequently placed under interdict and Churches were closed. By 1215, the continuous years of unsuccessful foreign policies adversely affected the barons. They united, rebelled against King John and forced him to sign the *Magna Carta.*

The *Magna Carta* consisted of a preamble and 63 clauses that imposed limitations on the king. Clause 39 became a significant development for the rule of law in England and later for mankind. It stated that "no free man shall be arrested or imprisoned or dispossessed or outlawed or exiled or in any way victimized…except by the lawful judgment of his peers or by the law of the land."

During this period, of equal importance were the changes to the Great Councils of Barons summoned by the kings. Generally, kings were only called to gain approval for new taxes, with little discussion on other matters that affected them. In 1258, the barons further imposed, this time on King Henry III (1207–1272) born in Westminster, the *Provisions of Oxford*. Kings were now required to govern in consultation with their subjects. The Great Councils, renamed Parliament, were to be assembled three times a year. Their composition was fluid, being divided into Commons and Lords. At this early date, England thus developed some of the first instruments for constraining royal power.

Marco Polo

In the thirteenth century, the grass lands of the Asian steppes were conquered by the Mongols under the leadership of Genghis Khan (1162–1227). He established the Mongolian Empire (1206–1368), which became the largest contiguous land won in history. It consisted of the combined conquests of Genghis as well as that of his descendants. The Mongolian Empire stretched from the Pacific Ocean, through the Asian steppes, to the Danube River and Persian Gulf in the west and united the tribes of the vast Eurasian plateau, including parts of Russia and China. For a century, the Mongolian khanates ensured a safe passage for traders across Asia, encouraging commerce.

It was during this time, that Marco Polo, together with his father and uncle, left Venice as merchants for China in 1271. Both elders had traveled there previously. After four years of arduous travel, the group reached China and the Mongolian ruler Kublai Khan (1215–1294). Kublai Khan, the great grandson of Genghis Khan, was both a statesman and a general who had founded the Yuan Dynasty (1271–1368) in China. The Polos intended to be away from Venice for just a few years; however, Kublai Khan's acceptance of the Polos offered the foreigners exceptional access to his empire and wealth.

Kublai Kahn employed the Polos for numerous official and diplomatic endeavors. After 17 years in his service, the Polos were able to negotiate their return trip to Venice. Their return path used a sea route to the Persian port of Hormuz, with the remainder of their travels through Turkey. After their lengthy return, Marco Polo commanded a war ship, which was captured by the Genoese, and sentenced to a Genoese prison. It was there he had the time to write his story. The book established Marco Polo as a celebrity and was published in Latin, Italian and French. After his release from prison in 1299, Marco Polo returned to Venice, married, and for 25 years continued the family business. His account of his experiences in China and Mongolia provided the West with its first image of this unique world. Polo's book was viewed by many, however, as fantasy. For example, paper money surely did not replace coinage. The public's fascination illustrated how remote and unknown the lands of the Far East still were to the Europeans.

Late Middle Ages: 1300–1500 (Challenges)

Avignon France

Conflict between the church and state continued well into the fourteenth century, this time initiated by the growing power of the Kingdom of France. Conflicts erupted over who had the superior authority, the pope or a monarch. It also involved who could tax the clergy and the questioned of disbanding the Knights Templar.

The Knights Templar, founded in 1119, consisted of both monks and knights who answered only to the papistry and was designed to aid pilgrims traveling to the Holy Lands. Eventually, the Knights Templar became wealthy and the French king became interested in seizing their French property. Eventually, Pope Clement V (1264–1314) acquiesced circa 1312 and ordered the dissolution of the Templars throughout Europe. He was the pope who also moved the Papacy from Rome to Avignon, France in 1309, where it remained until 1377.

Outside of France, pressure increased to return the pope to Italy. Papal authority rested on the fact that the bishops of Rome, namely popes, were the inheritors of Saint Peter. In 1378 several papal elections were held, during which time two popes were elected, Urban VI who remained in Rome and Clement VII who returned to Avignon, under the shadow of royal French power. Neither pope was willing to give up his authority. The discord lasted another 40 years, with even a third bishop claiming to be elected a pope. This time was known as the Great Papal Schism (1378–1417).

Numerous papal councils were formed to resolve the issues but they split along nationalistic lines. After 40 years, the Council of Constance was summoned by the Holy Roman Emperor. This council precipitated a single Roman, Pope Martin V (1369–1431) in 1417. Later, Martin V stressed the leadership of the papacy stating in a papal bull, that no one could again appeal a papal ruling to a council. Eventually, the Great Papal Schism weakened confidence in the leadership of the Catholic Church, which contributed to the Reformation.

The 100 Years War

The Hundred Years War (1337–1453) was the result of prior struggles between England and France that was embedded in land ownership issues, arising from the Norman conquest of 1066. The war erupted in 1337 over a relatively minor feudal issue and then escalated into whether or not the Kingdom of France would remain free of English control. The war continued until 1453 when the French regained the bulk of her possessions, which the English had previously claimed.

During this intermittent war, major changes were introduced in how the two kingdoms gained and organized the resources needed to pay for war. In the process, the ruling structures in both countries were enhanced. Prior to the war, the essential projectile weapon was either the short bow, which could not penetrate a good suit of armor, or the cross bow, which could but had a slow rate of reloading. In contrast, the long bow, which the English began using effectively in 1346, could penetrate armor. It had both the power of the crossbow and the repetition speed of the short bow. Accordingly, kings began to rely on inexpensive foot soldiers in an infantry with longbows.

Preceding the war, kings could and did impose kingdom-wide taxes only in emergency cases. They were obliged to pay for the routine issues that arrived, either from indirect taxes or from their own resources. As the war dragged on, however, emergency became routine and the collection of kingdom-wide taxes became a ritual that lasted even after peace was obtained. As a result, kings began using the extra resources to finance bigger armies that relied to a greater extent on foot soldiers. The extra resources gave kings greater financial reach than the barons. Royal power relative to the barons increased. As a result, in 1450, the French created the first standing army since the Roman Empire.

The Black Death

In 1000, the European population began expanding. By the end of the High Middle Ages, Europe was becoming overpopulated. Marginal land was used to feed the increasing population. During the late thirteenth and fourteenth centuries the Medieval Warm Period ended. The climate began to cool and became known as the Little Ice Age. As the weather became more erratic, harvests failed and famines occurred. The Great Famine of 1313–1317 arrived in Northern Europe and spread throughout Europe. Millions perished, malnutrition and disease were endemic. Recovery was delayed until 1322.

After the Great Famine, the Black Death (bubonic plague) occurred. It first appeared in Mongolia 15–20 years earlier. By the start of 1347, it reached the Genoese trading towns of Tana, at the mouth of the Don River, and Caffa in the Crimea. The epidemic was then carried to Trebizond on the Black Sea and to Constantinople by ship. Following the shipping lanes, the Black Death was in Alexander, Venice and Genoa before the end of 1347. By 1348, it had spread to many of the cities of Italy, France, Spain and then along the sea routes to Britain and Ireland. It continued for another five years, circling eastward through Germany and Sweden in 1350, Poland in 1351, and eventually Russia in 1352 and 1353.

Within years, 20 million people died. Mortality rates were estimated to be one-third to one-half of the population. The weakened health of the population, prior to the Black Death, contributed to the extreme loss of life. By 1450, the population of Europe had decreased 60% below that of 1300. This depopulation affected all regions: cities, towns and rural areas. For instance, Florence's population of 120,000 in 1338 decreased to 38,000 by 1347 and some villages were abandoned entirely.

Despite this loss of life, medieval European society continued to function but the social ordering was modified. So many laborers died that workers were in demand by landowners. This increased wages and the purchasing power of the poor increased.

Simultaneously, owners of the great estates also died, which resulted in large land holdings going on the market. Many heirs, strapped for income, sold acreage to peasants who previously could never have owned property. As a result, economic inequality decreased in much of Europe. In northwestern Italy, the share of the wealth owned by the richest 10% dropped from 61% in 1300 to 47% in 1450. In the same region, estate owners increased the wages of the workers, which contributed to the growth of a middle class. In contrast, the nobility in southern Italy enacted laws to prevent peasants from abandoning their estates in order to bargain for better jobs. Historians have suggested that the division of Italy into a poor south and a rich north resulted from these differing policies.

In other parts of Europe, many landowners and other employers in towns and cities tried to freeze wages to pre-plague levels or to revive serfdom. Nevertheless, most efforts failed due to a strong peasant resistance and lack of royal support. Eventually, those in power learned that they were accountable to every level of a community. This transition helped usher in the coming Renaissance. In the east, however, serfdom, which had been largely absent, grew in the large landed estates. In Russia, serfdom continued until its Emancipation Reform of 1861.

Gunpowder

Gunpowder was first discovered in China circa 850, during a search for the elixir of life. The mixture of 75 parts of saltpeter (potassium nitrate) with 15 parts charcoal and 10 parts sulfur did not enhance life but it did explode when lit by a flame. Thus, gunpowder was born.

Knowledge of this explosive substance was transferred along the silk routes to India, the Middle East and eventually to Europe by the middle half of the thirteenth century. As with the other Chinese inventions of silk, paper and the magnetic compass that found their way to Europe by similar routes; gunpowder had perhaps the most profound impact on human history.

The Chinese used gunpowder as an explosive. The Europeans, in contrast, began using it as a propellent. The Italians were the first to employ it in cannons with stone cannon balls in the early fourteenth century. By the late fourteenth century, the cannon technically improved to the point of being indispensable in any siege. By 1430, stone cannon projectiles improved to cast-iron cannon balls. The corning, drying and compression into cakes, of gunpowder in the fifteenth century, first accomplished by the French, increased the efficiency of the detonation and enabled gunpowder to be stored indefinitely. The arquebus, a primitive type of firearm with triggers, was in widespread use by 1500. In 1525, the battle of Pavia demonstrated its effectiveness against the knights.

This new technology changed the armor knights wore and the shape of castle walls. The popular chain mail was replaced by plate armor with beveled surfaces designed to deflect projectiles but was heavy and awkward to use. The high thin walls of medieval castles were effective repelling soldiers for centuries but were ineffective against cannonballs. In response, castles and towns made their walls shorter and thicker, with sloping surfaces to deflect the incoming projectiles. The impregnable land wall of Constantinople, which had been unassailable for nearly a 1000 years, was an excellent example of the power of the newly developed cannon. The Ottoman Turks used the new cannons to end the Eastern Roman Empire in 1453.

The development of weapons with gunpowder, as well as the longbow and the pike reduced the effectiveness and need for knights. Together with the increased costs of running big estates, the nobility needed to look for new employment. Many were attracted to the standing armies that began to emerge. Others established themselves in the courts of other nobles with greater resources or in the royal courts of kings. In the process, the ethos of the chivalric knight was dissolved and transformed into that of the courtier. By 1500, the knight's code of conduct with its jousting tournaments became archaic.

Printing Press

In the first century BCE, the hand written book was developed by the Romans (the codex) replacing the scroll, both of which used papyrus or parchment. In contrast to the scroll, the codex was made of uniform size sheets, bound together along one edge, in between two larger safeguarding covers. Large amounts of written information could be concentrated into a single easily transportable tome for the first time. A few centuries later, paper was developed by the Chinese, which replaced the expensive parchment. By the eleventh century, paper was brought to Europe, and by the thirteenth century, the Spanish were making paper in mills, using waterwheels, which crushed the plant and textile material used in the paper.

Handwriting was labor intensive. It could be replaced by repetitive printing, if available. The oldest known printed book originated in China circa 868 CE, using a technique of block printing that employed panels of hand-sculpted wooded blocks in reverse. Still, it too was a slow process. With paper, all the ingredients were available for mass producing information; although this did not happen until development of the printing press in 1450 by the German goldsmith, Johannes Gutenberg (1400–1468).

As lay literacy increased it created an interest and demand for books. The printing press satisfied this need, with the relatively inexpensive use of paper. Gutenberg replaced wooden blocks with metal type that was movable and durable and could be positioned easily and quickly in a variety of ways. He also developed an efficient pressing mechanism to apply the inked letters to the paper, based on his familiarity with wine presses. The first book that Gutenberg printed with his movable type was the Bible, in 1455 in Latin.

In contrast to Asia, printing spread rapidly in Europe because of the simple Roman alphabet. By the 1470 s, German printers were setting up printing presses wherever there was a demand for books. According to one estimate, by 1500, printing presses were in 236 towns and had printed over 20 million books. The printing press increased and quickened the spread of knowledge. In comparison to limited handwritten texts (often with only one version known), the printing press made the foundation of knowledge more secure with the increased numbers of books produced. Errors introduced by the printers were easily rectified in subsequent editions. Collective learning took a quantum leap forward.

Italian Renaissance

Background

Italy remained fractured since the demise of the Roman Empire. The memory of having been conquerors of the world, however, never vanished as its ruins were a constant reminder. Urbanism, literacy and trading at no time fully disappeared. Most of Italy spoke some dialect of Italian, which united them, as did the obedience to the bishop of Rome.

By 1400, in the south of Italy were the kingdoms of Sicily and Naples. The Papal State occupied the center, stretching northward on the easterly side of Italy toward Bologna and Ravenna. North of Ravenna was the Republic of Venice, with the further north controlled by the German Empire. Within these major structures (the church and Germans), the cities were the centers of power, culture and wealth. They either had allegiances with the church and were referred to as Guelph or held ties to the German Empire and were referred to as Ghibelline. By playing one against the other, cities could maintain their relative independence.

In the center of Italy, lay Tuscany, a hilly region ideal for the fostering of small independent city-states, mostly republican, that tried to maintain their independence while expanding at the expense of their neighbors. The city of Florence, the cradle of humanism (human values and experience) and the Renaissance, began as one of those early republican city-states.

Earlier, the Crusades had an evolutionary effect on Italy. The need for supplying transportation, equipment, supplies and loans to tens of thousands of knights, crossing the Mediterranean Sea from the Italian ports to the Holly Land immediately benefited the ports directly involved, i.e. Venice, Pisa or Genoa. The multitude of technical services and additional manpower needed were filled by people from the interior. This new economy changed the social and political structures of the towns. A broadly formed administration of towns based on men of influence, wealth or learning emerged. By 1200, the flexibility and independence of the evolving economic activities generated by the Crusades fueled wealth in these growing cities and helped to establish the preconditions for humanism.

The leaders in these city-states had little interest in the agrarian feudal past, where warfare was disruptive to commerce. In the evolution of Rome, they saw substantial evidence that the knowledge the classical age created. Rome was urbanized, secular, cosmopolitan, republican and educated. These influential men, educated with classical ideals, observed the relevance of the past to their communities. Instead of competing in war, eventually these communities competed for grandeur.

In contrast to the trade practices that restrained northern European commerce, individual Italian cities became freer than when ruled by an emperor or pope. The ban on usury was generally avoided. Royal taxation was lower, as political support from the affected city was desired. Restraints imposed by the protective guilds in the north were less. As a result, mercantile and political flexibility to control major capital accumulations enabled a revolution in trade and commerce. When new opportunities required capital greater than an individual merchant could muster or when the complexities of trade necessitated a more shared responsibility, new forms of business associations evolved.

Florence

Florence was an early influence in this new form of industrial manufacturing. Merchants controlled the import of raw wool, the production of wool and textile processes, and the export of the finished products. Comprehensive manufacturing of high-quality woolen cloth became a source of great wealth. The papistry contributed to its wealth as well. Because Florence was a leading pro-Guelph state, and had the necessary connections throughout Europe for resources and expertise, it became the collector of papal dues. This function resulted in substantial profits, during the transfer of dues back to Rome. The stability of the gold florin, after its introduction in 1252, further reinforced Florence as a city of international trade that could open new ventures for making money.

These innovations in business generated a new class of citizen in Florence and throughout northern Italy. The merchant patricians were lay, urban, cosmopolitan, educated, powerful and wealthy. Their skills in languages, accounting, law and mathematics aided business and added to Florence's wealth. They found the ideals of the humanist Petrarch (1304–1374) and Florentine humanist statesmen appealing. In Florence, the merchant patricians became the vanguard of the Renaissance with the bankers and industrialists, such as the Medici. They dominated the government socially, culturally and economically, and turned Florence into a city of art.

Florence's two iconic structures illustrate the early grandeur of the city: the Ponte Vecchio, which bridged the Arno River in 1345, and the Cathedral di Santa del Fiore. The cathedral began in 1296 and was completed with its massive dome by Brunelleschi in 1436. The cathedral's elegant, eight-sided dome was an engineering marvel: its 44 m diameter had no buttresses to hold it up.

Petrarch

Petrarch (1304–1374) saw little value in the study of scholastic methods in which arguments over minute points of faith appeared to have little relevance to how to live life. Instead, he was concerned with just and moral human behavior and believed that these principles could be achieved by a study of the literature of the great masters of classical antiquity. These masters had direct experience of life and human nature and provided insights into the results of good and poor behavior. He saw Cicero (106–43 BCE), for example, as not just as a lawyer, politician or statesman, but as a moral philosopher. Petrarch felt that from the study of the eloquence of past Latin writings comes truth and that good style, good writing, good speech leads to good and fair social values.

Thus, the study of the ancient past was reborn. It filled the schools of humanists, giving birth to the Renaissance. Petrarch's writing spread throughout Italy and to the collective conscience of western Europe. It promoted an emphasis on moral philosophy at the expense of scholastic reasoning and theology.

Humanism

The students of the humanist teachers were men not destined for the church, as they had been in the Middle Ages, but for future merchants, bankers, politicians, ambassadors and secretaries. Occasionally gentlemen had need of skills in arms and fighting. So, the education included a man's body and endeavored to develop it along with the mind, instead of denial of human flesh.

Students were encouraged to excel individually within shared values, attitudes, beliefs and morality and to include striving for glory, fame and recognition for posterity. Morality included a study of ethics, right and wrong, justice and loyalty, all of which were derived from studies of literature in both Latin and Greek. History was universally taught. Roman history was seen by the Italians as their own history. Historians such as Herodotus and Thucydides provided good examples of the practices that could support the creation of great states or, alternatively, the vices and practices that could later ruin them. What was important was the style and extent of moralizing.

Authors, such as, Cicero, Caesar, Livy (59 BCE-17 CE) and Plutarch (45–120) were popular. Some geography (Strabo [64 BCE-24 CE]) and astronomy (Ptolemy) were taught; however, a coherent approach to the natural world was generally absent except for a few enlightened teachers. Rhetoric was as central to the humanist education as it was throughout Roman times. It established a procedure of influencing an audience during meetings, council assemblies and public ceremonies. Its methods helped persuade audiences with sound advice and reasons for moral behavior.

Arts

During the Renaissance a parallel development occurred in the arts (painting, sculpture and architecture), embracing the rediscovered ways of thinking and viewing the world. Prior to this period, the painter Giotto (1280–1337) had already, using shading, represented figures with volume, substance and emotion. With his and Roman influence, a new artistic world was created. Together with three-dimensional linear perspective, painting could accurately reproduce what the eye saw. The landscape was recreated in realistic detail.

The humanist's desire to know the human condition was transferred to understanding the human body. Portraiture illuminated what an individual actually looked like, often showing emotion. The nude body was honored and became a celebration of God's greatest creation. Artists became respected and signed their work, in contrast to those of the Middle Age, where the creators remained anomalous.

Below are a few of these Renaissance legends:

Donatello (1386–1466) sculptor, celebrated for his realistic figures.

Sandro Botticelli (1445–1510) painter, known for the *Birth of Venus*.

Leonardo da Vinci (1452–1519) inventor, architect, painter of the *Mona Lisa* and the *Last Supper.*

Niccolo Machiavelli (1469–1527) philosopher, diplomat, writer, author of *The Prince.*

Michelangelo (1475–1564) architect, painter and sculptor, carved the *David* and painted the Sistine Chapel.

Raphael (1483–1520) student of da Vinci and Michelangelo, painter of the *Madonna* and *School of Athens*

Titian (1488–1576) painter of portraits.

University of Padua

In 1222, scholars from the University of Bologna, were looking for more freedom from town authorities, established the University of Padua as an international school. Students were partitioned into different groups called, Nations, according to their country or geographical area of origin. A representative was elected from each Nation. The representatives, elected a University Rector. The curriculum and selection of the teachers were administered by the Rector. The University was controlled by the students, which led to open discussions and the extensive dissemination of new ideas and philosophies.

When the students had completed their education and returned home, it resulted in the transmission of the newly discovered concepts throughout the rest of Europe. The University of Padua became a vanguard for the dissemination of humanist thought and culture. During the Renaissance, it gained a reputation for advances in the fields of law, art and medicine.

In 1405, Padua was conquered by Venice, which gave it the protection of the Venetian Empire. This enabled the university to benefit from exceptional academic freedom at a time when most universities were run by ecclesiastical organizations. The Venetian senate gave Padua University a secular status by being controlled by the Venetian State instead of the Archbishop of Padua. The university was designated studium venetum, which gave it special privileges. Oaths of fidelity to the Roman Catholic Church were not required. Students were not prevented from attending the University for religious reasons.

During this time, the University of Padua flourished, while others like that of the University of Bologna, which was controlled by the church, declined. The following are a few examples of the students and teachers at the University of Padua that illustrate its importance, during the High Renaissance:

Thomas Linacre (1460–1524) student, translator of Galen, founder of the English Royal College of Physicians.

Nicholas Copernicus (1473–1543) student, author, revolutionized astronomy writing *On the Revolution of the Heavenly Bodies* (1543), establishing the heliocentric universe.

Andreas Vesalius (1514–1564) teacher, anatomist, author of the revolutionary text-book on anatomy *De humani corporis fabrica.*

Galileo Galilei (1564–1642) teacher, confirmed Copernicus' model, changed the understanding of the cosmos.

William Harvey (1578–1657) student, discovered circulation of blood.

Decline

By the early seventeenth century, the energy of the Renaissance had dissipated as a result of many factors. French, Spanish and German invaders, fighting for territory within the Italian peninsula, created instabilities and disruptions. The discovery of new routes to the Orient, bypassing those of the Mediterranean Sea to northern Europe, led to a period of economic decline and limited the amount of money wealthy patrons spent on the arts.

Perhaps the greatest effect on the decline was the Catholic Church. The Church censored writers and artists in response to the Reformation. The Roman Inquisition was established in 1542 to judge what and who were orthodox or heretical. This judgment was reinforced by the *Index of Prohibited Books*, listing what Catholics could or could not read or even know. These events had a chilling effect on the dynamic energy of the Renaissance, restricting the ability of individuals to think independently or to question entrenched assumptions.

Humanism was stopped in its development. Faith replaced reason. Galileo, for example, who established that the Earth revolves around the Sun was deemed heretical by the Roman Catholic Church, prohibited from publishing his work and held under house arrest until his death in 1642. It was not until 1992 that the church pardoned him. Knowledge of the ancients, which was almost lost, was returned to life in the newly evolving Europe of exploration.

Byzantine Empire Ends

Between 1280 and 1324, the Ottomans who began as one of the Seljuk Baronies of Anatolia, conquered most of northwestern Turkish Anatolia. In 1345, they advanced into the Balkans and Greece and in 1361, captured the city of Adrainople where they established their capital. In less than 100 years after advancing into Eastern Europe, the Sultan Mohammad II (the Conqueror) lay siege to Constantinople with an army of 100,000 men just outside its great land wall. The triple-tiered wall and moat of 6.5 km long, had held off invaders for almost 1000 years.

However, this time the invaders had a new weapon, the cannon. The centerpiece of their weaponry was a large bronze cannon developed by the Hungarian engineer Urban. Initially, Urban offered his services to the Byzantine emperor but the emperor was unable to adequately pay. Mohammad II, could and did pay Urban. After 55 days of bombardment with cannons, the city walls were breached on May 29, 1453.

The Turks were lenient in dealing with the captured city and its population. The Ottoman capital was transferred to Constantinople. The city was repopulated and its great church, the Holy Wisdom (the Hagia Sophia), became a Mosque, symbolizing the transition from Christianity to Islam. Following the collapse of the western half of the Roman Empire, the eastern half was finally finished after another 1000 years of rule.

Western Europe was surprised by the collapse. The loss of Constantinople had longterm consequences. The buffer, preventing expansion of the Turks northward and west was gone. Between 1453 and 1550, the Ottomans tripled their territory. The Kingdom of Hungary became even more important in halting Turkish aggression to the north. In 1529, the Ottomans reached their maximum extent in Europe when they failed to capture Vienna by siege. Their expansion along the eastern Mediterranean Sea (including Egypt, much of Mesopotamia and Arabia), however, continued to grow. The entire southwestern rim of the Mediterranean Sea from Egypt to Gibraltar was now controlled by Muslim states.

The Muslim advancement appeared infinite. It was in this atmosphere of the feared Turkish expansion that precipitated the removal to push the last non-Christians out of Spain. In 1479, the dynastic union between Isabella of Castile (1451–1504) and Ferdinand of Aragon (1452–1516) created a new Christian dynamic capable of achieving this goal. In 1492, the last remaining Muslims, the Moors of Granada, were defeated. Jews were expelled along with Muslims, unless they converted to Christianity.

Draw of Spices and Gold

Another unintended consequence of the collapse of Constantinople and the expansion of the Ottomans into the Mediterranean Sea was the motivation to find other ways around the Muslim middle men. They were preventing Europeans from establishing direct economic contact with the Far East for spices and other luxury items but also with sub-Saharan west Africans for their gold. The Portuguese were ideally suited for initiating both possibilities by exploring new routes down the west side of Africa; and eventually to the Far East.

The Christian Portuguese had gained their independence from the Moors by the eleventh century and from the Castilians in 1386. They were a seafaring people familiar with the Atlantic Ocean. Their first forays down the African coast encouraged the development of the caravel, a ship that made long voyages practical along the African coast and into the Atlantic. The caravel used both square and triangular sails. The square sails provided speed and the lateen triangular sail, enabled tacking into the wind and provided greater maneuverability. The addition of naval cannons, mounted between the caravel's ribs and along its sides, gave protection from enemies.

The Portuguese, with the support of Prince Henry the Navigator (1394–1460), carefully explored farther and farther south along the west coast of Africa, accurately mapping their routes, coastline, currents and winds. Along the way trading posts were established. The Portuguese merchants exchanged weapons and textiles for gold, ivory, cotton and slaves. In 1434, the Portuguese went around Cape Bojador with its perilous shoals and winds, reaching the mouth of the Congo River in 1483. This was new territory, unknown to the Romans. By 1488, Bartholomew Diaz (1450–1500) reached the southern tip of Africa, the Cape of Good Hope, proving that Africa was finite and could be circumnavigated. Ten years later, Vasco da Gama (1460–1524) did just that.

Taking advantage of the anti-clockwise wind pattern of the south Atlantic, Vasco da Gama sailed significantly west of Africa before turning east and going around it. He arrived in Calicut, known as a City of Spices, on the southwest coast of India in 1498. On his return, he sailed home with a cargo of cinnamon and pepper. Vasco da Gama was followed by Alvares Cabral (1467–1520) and Afonso de Albuquerque (1453–1515). They were allowed to build a fort and church at Cochin located 169 km south of Calicut in 1500. The Muslim monopoly disintegrated.

The successful voyages of the Portuguese, however, adversely impacted Renaissance Italy. Christian merchants were able to bypass dealing with the Muslims and the dangers of the Mediterranean Sea. The price of spices and luxury goods from the east dropped precipitously. In 1505, spices could be bought in Portugal for 20% of those sold in Venice.

Chapter 8
Transition: 1500–1700 (The Known World Expands)

Voyages of Discovery

As the Mediterranean Sea was transitioning to Turkish control and routs around Africa were being developed, Europeans began looking westward for new investments. Several pivotal explorers with the courage and capability garnered funding for their ideas. Their discoveries changed how Europeans viewed the world, including places, cultures and people not known to exist and who were not mentioned in the Bible.

Christofer Columbus

Christofer Columbus (1451–1506) believed that by sailing west, the riches of the Indies could eventually be reached. He was an Italian born in Genoa, who moved to Portugal in 1476, where he learned to command his own ship and understand the circular wind patterns of the North Atlantic. Using the largest estimates of the Eurasian land mass and the smallest diameter of the Earth, Columbus calculated (incorrectly) that Asia was only 4800 km west of the European coast and reachable by ship.

Initially, Columbus tried without success to convince the King of Portugal to support his efforts to sail west. Failing that, in 1485 he spent the next seven years pursuing this ideas with Ferdinand (1452–1516) and Isabella (1451–1504) of Spain who were interested in tapping into the riches of the East. Eventually, the Kingdom of Granada was captured by the Spanish, money was made available and Columbus was given the ships he needed. On August 1492, with three ships and crews, Columbus set sail from Spain, first to the Canary Islands and then he followed the clockwise wind pattern blowing to the west.

© The Author(s), under exclusive license to Springer Nature Singapore Pte Ltd. 2024   115
T. Sanford, *A Whirlwind History of the Universe and Mankind*,
https://doi.org/10.1007/978-981-97-2674-5_8

Holding his fearful crews together on the open ocean, Columbus eventually reached landfall 33 days later in the Bahamas. After finding Cuba and Hispaniola (Dominican Republic and Haiti) in the Caribbean Sea and trading a few ornaments for gold, Columbus returned to Spain. He sailed back on a northerly route following the clockwise trade winds blowing east toward the Azores. On his return, he exaggerated the value of his discoveries, especially those of gold and spices, which disappointed future settlers.

In total, Columbus made four trips to the Caribbean, demonstrating that the Caribbean Sea was enclosed on the westerly side. His administration of the new territories was considered a failure. Columbus' talent lay with his navigational skills. Mistakenly, thinking that he had found a new route to the Indies, he called the peoples that he found Indians. At the time of his death, Columbus was still unaware that he had discovered a whole new continent.

Amerigo Vespucci and Martin Waldseemuller

Following in Columbus's path was Amerigo Vespucci (1454–1512), a Florentine merchant, navigator, explorer and cartographer. Vespucci made two voyages to the west: the first 1499–1500 on behalf of Spain and the second 1501–1502 for Portugal. In the first, he sailed to Guyana, then south, discovering the mouth of the Amazon River. On his second, he reached the coast of Brazil. Continuing southward, Vespucci sighted the bay of Rio de Janeiro and likely Rio de la Plata, traveling farther southward along the coast of Patagonia. On his return he became convinced that the new lands discovered by Columbus and himself were not part of Asia but that of a New World.

In 1502 and 1505, he published two booklets in which he stated that Brazil was part of a New World. The announcement motivated the German cartographer Martin Waldseemuller (1470–1521) to recognize Vespucci's accomplishments by using *America*, a Latinized version of Amerigo, for the first time in his 1507 map of the world. It demonstrated the Americas as a separate continent and included the data Vespucci had gathered on his two voyages. Other cartographers followed suit and by 1532 the name America was permanently ensconced on all world maps.

Waldseemuller also printed maps designed to be cutout and glued onto spheres forming globes. The Waldseemuller edition of Ptolemy's *Geography* together with a set of Waldseemuller's maps were added to the 1513 edition printed in Strasbourg. The collection was thought to be the first example of a modern-day Atlas. His maps were the first to show a separate Western Hemisphere from Eurasia. They represented a significant advance in the understanding of the true shape of the world.

Nunez de Balboa

Nunez de Balboa (1475–1519) was a Castilian conquistador, explorer and governor who contributed to the founding of the first stable settlement in South America, Santa Maria la Antigua del Darien, which was located on the easternmost Isthmus of Panama. By 1511, he became its acting governor. On September 1, 1513 he lead an expedition of 190 Spaniards and a few Indian guides across the dense jungles of the Isthmus. From a mountain peak on September 25, 1513 he sighted the present day Pacific Ocean, becoming the first European to see it. Balboa continued and reached the sea four days later. Standing knee deep in it, Balboa claimed the sea and all its shores for Spain. News of this sighting arrived after King Ferdinand of Spain had already authorized Pedrarias Davila to become the new governor of Darien. Years later, Davila who was reportedly jealous of Balboa, ordered him arrested on charges of treason and had him executed in 1519 before he had time to explore the lands surrounding his ocean discovery.

Ferdinand Magellan

Ferdinand Magellan (1480–1521) was a Portuguese explorer who was inspired by Columbus and Balboa. No European had yet succeeded in sailing west to the Spice Islands and he was determined to be the first. Having returned from the East Indies, as an experienced seaman and studied the latest charts, Magellan approached King Manuel I (1469–1521) of Portugal seeking support for a westward voyage to the Spice Islands. Continuously rebuffed by King Manuel, in 1517 he was permitted to leave for Spain where he sought the support of Charles I (1500–1558), the King of Spain and future Holy Roman Emperor Charles V (1519–1558). Charles I, hoping that this might yield a commercially useful trade route, independent of the Portuguese, approved the expedition and provided the bulk of the funding. Charles I was the grandson of Queen Isabella and King Ferdinand, who previously had financed Columbus.

On September 20, 1519, with a fleet of five ships, Magellan left Spain and reached the Canary Islands on September 26, where he took in supplies for Brazil. On board were 230 Spaniards, 40 Portuguese and supplies for two years. By December, they made landfall in Rio de Janeiro. Searching southward for the hoped-for passage west, they wintered for five months in a sheltered harbor at the Port of Saint Julian. During the winter, one of the ships exploring for the route ahead was shipwrecked in a storm, with the crew later rescued. On Easter Day 1520, Magellan quickly quelled a mutiny initiated by the greater numbers of Spaniards and by spring, the search for a passage west resumed. By October 21, 1520, Magellan entered the strait that gave passage to present day Straits of Magellan.

In the cold dangerous conditions of the straits, the crew and one ship deserted and returned to Spain, taking much needed supplies. The remaining three ships reached the Pacific Ocean by the end of November 1520, having taken 38 days to transverse the 592 km straits. Magellan and his crew thus became the first Europeans to sail into the western sea that Balboa saw from the Isthmus of Panama. The sea was named Mar Pacifico because its waters were calm relative to those of the Straits of Magellan.

After passing through the straits, Magellan anticipated a brief sail to Asia. The journey, however, took nearly four months to reach land, which occurred on March 6, 1521 when they arrived at the Island of Guam. Their stores of water and food were exhausted and at least thirty men had succumbed to starvation and scurvy.

Once their stores of food were replenished, they continued to the Philippine Island of Cebu, which was only 644 km from the Spice Islands. Magellan and forty of his crew were killed on April 27, 1521 in a skirmish with a hostile tribe on the island of Mactan. After Magellan's death, with too few crew, one of the damaged ships was burnt. The two remaining ships sailed on to the Moluccas, where they filled their stores with spices. One ship attempted unsuccessfully to return across the Pacific Ocean. The other ship, the Victoria, continued west under the command of Juan Sebastian Elcano (1476–1526), a Basque navigator, leaving the Islands in early 1522 for Spain. Crossing the Indian Ocean, they passed the Cape of Good Hope on May 19, 1522, successfully avoiding any Portuguese. On September 6, 1522, the Victoria returned to Spain with just 18 of the original 270 men surviving, thus completing the first circumnavigation of the Earth. The shipload of condiments collected, was greater than the cost of the original fleet, illustrating the value of spices,

Antonio Pigafetta (1491–1531), one of the survivors and Magellan's assistant, was a Venetian scholar and explorer who kept a detailed account of the expedition. Eventually, *Voyage Around the World by Magellan 1518–1521 a Book by Antonio Pigafetta* was published by 1550. It proved invaluable to Europe's understanding of the world.

Even though the Greeks knew the Earth was theoretically round, the circumnavigation proved empirically that the Earth was round and not flat, decisively discrediting the medieval theory. Magellan's expedition found the Pacific Ocean to be massive and discovered that the Earth was much larger than previously thought. European knowledge of geography immeasurably expanded. Magellan's expedition was regarded as "…the greatest achievement in seamanship." The voyage illustrated the sophisticated level of the European organizational, engineering and navigational skills of the period.

New Exchange Networks

The Columbian Exchange

The super continent Pangaea, where all the continents were once interconnected, began breaking apart 200 million years ago. They developed into the continents known today. The last ice-free corridor connecting Siberia and Alaska, the Beringia land bridge, which allowed the first humans to migrate to the Americas, was severed 10 to12 thousand years ago when the connecting ice sheets began to melt. After that time, the Americas developed separately from Eurasia, allowing differing ecosystems to evolve. Columbus and the migrations that followed, reversed that separation, forming what historians call the Columbian exchange.

Cattle, sheep, goats and pigs were transported to the Americas (the New World) for the first time, beginning the process of colonization. The horse, which was eliminated earlier by Paleolithic man, returned. As in the Old World, the horse had an enormous effect on native cultures, which relied on subsistence farming and foraging. New crops were introduced: wheat, rye, sugar cane, coffee beans and bananas. Flowing back to the Old World were the foods: potato, sweet potato, tomatoes, maize, squash and tobacco.

Pathogens were transferred also. The Americas introduced syphilis to the Europeans but the major flow of diseases spread toward the Americas. The main killers were smallpox, typhus and measles, for which the Europeans had developed some form of immunity but were new to the Americas. In sixteenth century Mesoamerica (the area extending from central Mexico to what is present day Honduras and Nicaragua), historians estimate that the indigenous population declined by 90–95% and, in the Andes, by 70%. The relative ease and rapidity with which the Spanish conquered native peoples of these regions was significantly aided by the introduction of these diseases. The resulting breakdown of their political, social and religious organizations facilitated new communities based on the Spanish models.

The First Global Exchange

Many limited exchange networks had operated for millennia between the Mediterranean region and the Far East via land and sea along the silk and spice trade routes. After 1350, new trade routes expanded eastward from Portugal, around Africa, across the Indian Ocean to India, China and the islands of southeast Asia; however, none of these involved the entire world. With the unexpected discovery of the Americas in 1492, trade routes developed westward across the Atlantic, between Spain and Veracruz Mexico, and along the west coast of South America. By 1565, trade routes between Lima and Panama and between Acapulco, Mexico and Manila, Philippines, linked the entire globe for the first time. The rivers within the new lands began to be explored and knowledge was acquired about the unknown territories between them.

In 1545, massive amounts of silver were discovered in the mines of Potosi, located high in the Bolivian Andes. In 1548, rich silver mines were discovered near Zacatecas, 800 km to the northwest of Mexico City. From these mines, silver was transported to Mexico City where it was minted into Spanish pesos. A portion of the silver pesos was transported to Spain where the Holy Roman Emperor, Charles V, and then his son Philip II (1527–1598) spent it on wars to control their vast empire. Ultimately, silver was funneled into Spain's north European bankers to finance war and trade with Indian Ocean merchants.

In the opposite direction, another portion of Spanish pesos crossed the Pacific to Manila, where it was used to buy Chinese silks, porcelains and other luxuries. The silver eventually arrived in China, which had a need for silver in its coinage. The price of silver in the Americas was low, because it was mined by slaves and it was plentiful. The price was high in China because the mines were depleted. This need for silver in China drove the world's first global system of exchange and became a financial network, with the Spanish peso becoming the first global currency.

The Atlantic Exchange

Sugar became an example of a new exchange network that crossed the Atlantic Ocean. In the late fifteenth century, sugar plantations were developed in the Canary and Madeira Islands. They provided prototypes for those to be developed in Brazil and the Caribbean Islands. By the sixteenth century, the Portuguese established plantations in Brazil, followed by the Dutch, French and British, who introduced sugar plantations to the Caribbean in the beginning of the seventeenth century.

The plantations required large numbers of inexpensive laborers that were supplied by an emerging slave trade from Africa. Across the Atlantic Ocean, a highly profitable system of triangular trade evolved. Europeans provided the initial capital demand for sugar. The African slave traders provided the slaves. (In the Caribbean most of the natives had been decimated by disease.) In exchange for the slaves, Europeans provided the slave traders with weapons, metal products, textiles and wines, and the traders transported the slaves across the Atlantic to the plantations. The plantations transported the sugar back to Europe for use as a sweetener and to be distilled into rum.

Investors, merchants, slavers and plantation owners all made large profits at the expense of the African natives who were traded as commodity. Their cheap labor made the plantation system feasible. The system linked Europe, Africa and the Americas together into a single exchange network. Eventually, the plantation system expanded to include tobacco and cotton.

The Reformation and Religious Wars

While new knowledge and opportunities were flowing into Europe on its western boundaries, changes were occurring within its interior. Crises within the Roman Catholic Church, relating to the widespread corruption were continuing. It reached a critical point in 1517 with the nailing of Ninety-Five Theses to the Castle Church door in Wittenberg, by the German monk Martin Luther (1483–1546).

The catalyst was the sale of indulgences in Luther's town of Wittenberg. These spiritual pardons could be bought for a price from a papal legate to excuse any sin committed. The renunciation of papal authority relating to the sale of the indulgences increased. Luther believed that faith alone was enough to arrive at salvation. Within the Ninety-Five Theses was support for dissolution of the monasteries and changes to other aspects of Catholic orthodoxy.

Throughout Germany, Luther's ideas spread rapidly in a movement known as the Reformation. The Ninety-Five Theses and his refusal to retract them, as demanded by the pope, led to his excommunication in 1521. Luther, however, had gained many supporters, in particular, Frederick III, the Elector of Saxony (1463–1525). Under Frederick's protection Luther had time to think and write, providing his new religion, Lutheranism, a coherent shape. He helped bring Christianity directly to the people by translating the Bible into German.

Initially, the German princes supported Lutheranism because it enhanced their wealth and power. All the taxes previously owed to Rome, the monasteries and the church would instead become theirs. Luther, needing the support of the nobility to ensure the religious transformation, asserted that church property and law should be administered by secular rulers. The pope was relieved of his authority and God was returned to the top of the religious pyramid. The advance of the Lutheran Reformation was rapid. By 1529, Germany was split 50–50 between Lutherans and Catholics. By 1540, the Scandinavian kingdoms supported Lutheranism. The previous unity and power of the Catholic Church was altered.

This breach encouraged other reformers, i.e. Ulrich Zwingli (1484–1531), who led the Swiss Reformation in Zurich. Zwingli, attacked ecclesiastical abuses and supported priestly marriage but disagreed with Luther on other issues of the new church doctrine.

The Frenchman John Calvin (1509–1564) was influenced by Zwingli. In 1536, Calvin found a home in Geneva where he provided a message that was strong and popular. In 1536, his book *Institutes of the Christian Religion* was published. Those who were confused about the spiritual world found the answers in Calvinism. He proselytized Calvinism into cells that could be fomented and transplanted anywhere. Good behavior was encouraged and by demonstration, individuals could convince themselves that they were among the saved.

Calvin's printed book, spread by missionaries, expanded to the rest of Europe and promoted his religion. By the time of his death, France, like Germany, was split in half between French Calvinists (Huguenots) and Catholics. In the early part of the sixteenth century, Calvinism had spread into the mercantile environment of the Low Countries (the costal region of northwestern Europe, consisting of Belgium, the Netherlands and Luxembourg). Businessmen and merchants liked the role of the laity in the congregations of the Calvinists. Calvin's theology had a major influence on the development of several new forms of the reformation: Presbyterianism in Scotland and Puritanism in England. All Christian religions that conformed to the principles established by Luther, Zwingli, Calvin, etc., as opposed to the Roman Catholic Church, were referred to as Protestantism.

Like the European continent, England was not free from the circulating religious discontent. This included the taxation flowing to Rome, the extravagance of the monasteries and the non-Christian behavior of the clerics. In 1527, King Henry VIII (1491–1547) divorced his wife to help produce a male heir to the throne and the pope excommunicated him. As a result, the Parliament dismantled the Roman Catholic Church in England and transferred its jurisdiction and properties to the English crown. Henry became the head of the Church of England in 1534 with its accumulated wealth and the pope lost his authority in England. Eventually, the Church of English transitioned into a form not unlike those of the Protestant Reformation on the European continent.

The Catholic Church reacted against the Protestant Reformation by forming the Council of Trent. The Council deliberated between 1545 and 1563 and played a major role in the Counter-Reformation. The Catholic doctrine and hierarchy with central authority of the pope was reaffirmed, as was the use of the Latin Bible. Heresy was to be held in check by the continuance of the Roman Inquisition. In 1559, the *Index of Prohibited Books* was written to control heretical ideas.

These restrictions significantly restrained the freedom of inquiry in the Catholic world. Prohibited books included publications by astronomers such as Johannes Kepler's *Epitome Astronomiae Copernicanae* or philosophers like Immanuel Kant's *Critique of Pure Reason*. The banning of these types of books encompassed the 17 volumes of the *Encyclopedie* whose chief editor was Denis Diderot, which appeared between 1751 and 1765. This ban was formally abolished in June 1966.

A new Jesuit Order provided the means by which the Catholic faith would be taught and increased. It's priests, traveling with the conquistadors, would live among the communities of the New World, spreading, the faith by missionary work, preaching and teaching. Europe was becoming partitioned by religion. The resulting confrontation sowed the seeds for war. Europe, which had been uniformly Roman Catholic for a millennium was now in chaos. The choice of a state religion led to a century of war and large loss of life.

One of the last major wars on the continent, the Thirty Years War (1618–1648), ended with the signing of the Treaty of Westphalia in 1648. The German principalities gained autonomy over their religious and political affairs, including legislation, defense and taxation. The treaty gave the Swiss their freedom from Austria and the independence of Netherlands from Spain. France acquired the bulk of Alsace-Lorraine, becoming the strongest kingdom in Europe.

The possibility of a Roman Catholic reconquest of Europe was lost. Protestantism on the continent remained. The European continent was transformed into the essential institutions of modern political structure, which included: territorial sovereignty, having a centralized government with a bureaucratic administration and standing armies, all in the form of a secular state.

England remained Protestant under Queen Elizabeth (1533–1603). Her tolerance for the minority of Catholics changed, when the pope declared her to be a heretic and not to be obeyed by any of her Catholic subjects. Spain, under Philip II (1527–1598), with the blessing and aid of the pope, sent the Spanish Armada in 1588 to invade England and restore Catholicism. The Spanish invasion failed, securing Elizabeth's position on the throne.

As on the continent, religious reform continued to smolder and led to civil war in 1641. Eventually the Puritan, Oliver Cromwell (1599–1658), took over and ran England as a dictator until his death in 1658. After Cromwell, Parliament brought

Charles II (1630–1685) to rule England from 1660 as a monarchy until his death in 1685. Charles II was succeeded by this brother James II (1633–1701), who reigned in England, Ireland and Scotland until 1688. Discontent with James's pro-Catholicism resulted in Parliament asking William of Orange (1650–1702), from Holland, and his wife Mary (1662–1694) to assume the throne in 1688.

This peaceful transfer of power permanently established Parliament as the ruling power of England. The new Protestant regime benefited by the support of the majority of the country and ensured the end of civil wars. Supplementing this change, Parliament decreed that no Catholic would ever again assume the English crown.

While these political and religious structural changes were underway, the exploration of the New World continued. The Spanish and Portuguese claims to legal control of the New World were lost. During the early part of the seventeenth century, the first successful colonies on the North Atlantic coast were being established by the English, Dutch and French. These countries also began to form colonies in the Caribbean as well: St Kitts and Barbados by the English in 1624–1627, Curacao by the Dutch in 1634 and Guadeloupe and Martinique by the French in 1635.

The Scientific Revolution

Concurrent with the voyages of discovery, the new exchange networks, the Reformation and the religious wars, the Scientific Revolution in Europe was underway. In one scientific area, it manifested itself in attempting to understand the true position and mechanisms behind the movement of the planets. In solving this planetary enigma, the arrival of the Enlightenment Period was expedited.

Understanding this motion of the planets provides an ideal illustration of the Scientific Method. Its stages from hypotheses to measurements to explanation of the paradox occurred on a grand scale. These three stages, were spread over millennia among the many scientists involved.

Scientific Method

Hypotheses

During the Hellenistic era, Aristotle (384–322 BCE) imagined an Earth-centered universe with all astronomical objects positioned on perfect spheres, following symmetrical circular orbits. These were the governing principles of the heavens. Later, Aristarchus of Samos (310–230 BCE), a Greek astronomer, proposed that the Earth revolved around the Sun, in a heliocentric system, with the moon rotating about the Earth. His observations did not survive but his work was known by Archimedes (288–212 BCE) and others.

Subsequently, the Greek-Egyptian Claudius Ptolemy (100–170 AD) rejected all heliocentric explanations. He proposed a geocentric system to accommodate Aristotle's perspective, in which the Earth was the center of the universe and the planets revolved around it. To account for the imperfections from circular motion that were observed, Ptolemy postulated that their irregular movements resulted from a combination of smaller circular motions, referred to as epicycles, which were centered about the main circular orbit. His mathematical model, published in the *Almagest*, with the Earth as the center of the universe, became a cornerstone of western understanding of the universe and a canon of the Roman Catholic Church for the next 1500 years.

During the Renaissance, Nicolaus Copernicus (1473–1543), the Polish mathematician and astronomer formulated a mathematical model of the universe that placed the Sun at the center of the universe rather than the Earth. Copernicus also proposed that the Earth rotated on its axis, giving rise to the appearance that the Sun, planets and stars orbited our own planet. He was likely inspired by the thinking of Aristarchus. Although the Ptolemaic model provided a reasonable description of the data, the Copernican explanation, which did not agree exactly with perfect circular trajectories, was vastly simpler.

Copernicus finished his work *De revolutionizes erbium coelestium libra vi* circa 1530 but delayed its publication until 1543, the year of his death to avoid confrontation with the church. This delay was rational. In February 1600, the philosopher, mathematician and cosmological Italian thinker, Giordano Bruno (1548–1600), was burned at the stake in Rome for thinking such thoughts.

Measurements

Between 1576 and 1580, the Danish astronomer Tycho Brahe (1546–1601) developed precise astronomical instruments for measuring the positions of the planets and stars, which paved the way for future discoveries. His observations, which included a comprehensive study of the solar system and encompassed over 1000 stars, were the most accurate possible prior to the invention of the telescope. These measurements corrected the existing tables, like those of Copernicus. On November 11, 1572, Brahe saw a new star in the constellation of Cassiopeia, which was brighter than Venus, and lasted until March 1573. His careful measurements were published as *De nova stella* 1573. This observation (a supernova) was disquieting to a world that required the heavens, unlike the Earth, to be perfect and unchanging. This discovery, along with reports of the Copernican Theory that the Earth may not be the center of the universe, contradicted the unquestioned laws of antiquity, including the Bible.

The German astronomer Johannes Kepler (1571–1630) was Tycho Brahe's pupil. Together, the two created a new star catalogue based on Brahe's own precision measurements. This catalogue became the *Rudolphine Tables,* published by Kepler in 1627. At Brahe's passing, his data was left to Kepler, who continued to process it. The data demonstrated the heliocentric nature of the solar system, which Kepler was able to quantify into three mathematical laws of the planetary orbits: (1) the planets move in elliptical orbits with the Sun at one focus, (2) a line connecting the Sun and a planet sweeps out equal areas in equal times and (3) the square of a planet's orbital period is proportional to its mean distance from the Sun cubed. Kepler published the first two laws in 1609 and the third in 1619. Ptolemy's model was inconsistent with the precise measurements of Brahe and the analysis of Kepler. Ptolemy's model could have been discarded then and there.

In 1564, Galileo Galilei (1564–1642) was born in Pisa. He was a physicist, mathematician, philosopher, professor and astronomer whose measurements in 1609 also supported the heliocentric planetary motion. In that year, he was made aware of a new instrument invented by Hans Lippershey (1570–1619), a Dutch eyeglass maker, that could magnify objects by three times. Advancing on this achievement, by trial and error, Galileo discovered the secret of the invention, making his own magnifying telescope.

Galileo rapidly improved the instrument's magnification to 20 times when he began observing the heavens in the fall of 1609. Looking at the phases of the moon, he observed that the lunar surface was not smooth, as was thought, but rough, uneven and filled with craters. In January 1610, he discovered Jupiter with its four moons revolving around the planet, as if in its own small planetary system. Galileo also observed the phases of Venus, demonstrating that it, too, was unambiguously revolving around the Sun; images in the telescope showed the existence of many more stars than had been visible to the naked eye, and that the Sun, also had blemishes, sunspots that rotated in a circular motion. In March 1610 he quickly published these results in *Sidereus Nuncius,*.

In April 1610, Kepler publicly supported Galileo's findings and by August 1610, Kepler published his independent confirmation of Galileo's observations. The ensemble of this different kind of data, which only a heliocentric system could explain, revolutionized astronomy and eventually led to acceptance of the Copernican model.

Although the Copernican model was confirmed by measurement, the laws of Kepler only provided a kinematic description of the orbits. Another hundred years passed before Isaac Newton (1642–1727), the English physicist, mathematician and astronomer, developed a dynamic explanation of the orbital motion. Kepler's measurements and analysis, however, were a major step in answering the question, of identifying the true positions and movements of the planets. Together with the earlier hypotheses and observations, the first parts of the Scientific Method were illustrated and assisted in establishing the Scientific Revolution.

This model contradicted writings in the Bible and had the potential to undermine the authority of the Catholic Church. The church had based much of its doctrine on the geocentric Aristotelian belief in the immutability of the Heavens. Eventually, Galileo was brought before the Roman Inquisition in 1633 and given a choice: be burned at the stake or recant and spend the rest of life under house arrest. He chose the latter.

In a letter Galileo wrote to Benedetto Castelli (1578–1643), an Italian Benedictine student and friend of Galileo, Galileo argued that "the Bible was an inspired text, yet two truths could not contradict one another. So in the case where it was known that science had achieved a true result, the Bible aught to be interpreted in such a way that makes it compatible with this truth. The Bible was a historical document written for common people at a historical time and had to be written in language that would make sense to them and lead them towards the true religion." (Quote taken from Stanford Encyclopedia of Philosophy...Galileo Galilei, May 10, 2017). If the Church had followed Galileo's advice, many future lives would have been saved.

During his confinement, Galileo's contributions to science continued. Prior to his significant astronomical discoveries he was developing his own Laws of Motion. These were based on measuring the trajectories of objects on smooth or frictionless surfaces, rolled down or up inclined planes, dropped from heights or attached to pendulums. From these careful experiments, he evolved (1) the concept that the motion of a body has both direction and speed, namely velocity, (2) the concept of a force to be the cause of motion, (3) the natural state of a body was rest or uniform motion and (4) a body resists a change in motion, inertia, which he discovered by isolating and removing friction.

Galileo also developed the Law of Falling Bodies, which states that all bodies fall at the same rate of speed, acceleration, independent of weight. This Law directly contradicted the Aristotelian view that heavy bodies fall faster than lighter ones. He generated a mathematical expression demonstrating that the distance fallen is proportional to square of the time fallen. He determined that the trajectory of a projectile was that of a parabola due to the combined motions of uniform acceleration downward and constant speed. He established that only those motions differing from a shared common motion could be understood as moving. This statement became his principle of Relativity of Observed Motion.

During house arrest, Galileo finished his studies on motion and his observations on the strength of materials, which had been interrupted by his astronomical work. That work was able to be smuggled out of Italy and published in Leiden, Netherlands in 1638 under the title *Discorsi e dimostrazioni matematiche intrno a due nuove scienze attenenti alla mecccanica.* This distillation of motion became even more significant than his astronomical work.

Galileo felt that these mathematical principles of matter and motion belonged to the new science of mechanics, the name he gave to this way of thinking. He became one of the early philosophers, similar to Pythagoras, to state that the "laws of nature are written in the language of mathematics." As the result of his pioneering studies, Galileo is considered to be the father of science. Science comes from the Latin word scientia, meaning knowledge.

Explanation

The observations of Galileo lay the foundation for the English physicist and mathematician Isaac Newton (1642–1727) insights, who was born the year of Galileo's passing. Newton used and expanded Galileo's work and clarified the mechanism behind the planetary behavior, providing insight into the final chapter of the Scientific Method illustrated here.

Newton's initial three Laws of Motion and Force are: (1) a body at rest or in uniform motion remains the same unless acted upon by a force, (2) a force F acting upon a body of mass m generates a resulting acceleration a determined by the equation $F = m\,a$, and (3) if body B exerts a force upon body A of F_{ab} then body A generates an equal and opposite force upon body B of $-F_{ab}$. The bold quantities are vectors, which symbolize direction. The first law was that of Galileo and was known as the principle of inertia. The second law was called the equation of motion and was powerful. Given the mass of a body and the force acting upon it, its subsequent position could be determined precisely. The third law led to the conservation of momentum, mv, where v was the velocity of mass m.

Using these and Kepler's Laws of Planetary Motion, Newton discerned that the acceleration vector (of the force pulling the planet toward the Sun) always pointed toward the Sun and that the magnitude of the acceleration of a planet towards the Sun was inversely proportional to the square of the distance between the planet and the Sun, and not just the mass of the planet. Thus, the force holding the solar system together must be coming from the Sun. In similar thinking, Newton realized that the Earth exerts a weaker force on the Moon. And, the force pulling the falling apple from the tree must be the same force. This force, Newton surmised, must be an attractive force produced by the mass itself.

Newton was, accordingly, led to his Universal Law of Gravity, where the magnitude of the force F, now called gravity, between a body of mass m_a and a body b of mass m_b could be expressed by the inverse square of the distance r between the two masses as: $F = g\,m_a\,m_b/r^2$. The gravitational constant g was determined from experiment. The force of gravity was symmetric between the two masses and was the reason why it was referred to as universal; the mass did not depend upon whether or not it was a falling apple or a rotating planet.

With these laws, Newton was able to calculate the shape of the elliptical orbits and all of Kepler's phenomenological Laws of Motion from first principles. Newton initially had to develop a new mathematics called calculus, which permitted the equations to be solved. His work was first published in *Philosophiae Naturalis Principia Mathematica* (The Principia) in July 5, 1687, with a number of editions following until 1726.

With Newton's theory, humankind had a rational explanation for why the planets and Earth move as they do. The concept of unalterable forces and Laws of Nature was established. Once the universe was put into motion, Newton's laws kept it in motion. Conceptually, God was only needed to start the process. The discovery of the answer involved the iteration of hypotheses, improved instrumentation and connecting the measurements to a mathematical expression that agreed with all the data.

As a illustration of the power of Newton's equations, equating the second Law of Motion of a falling mass m, $F = m\,a$, with the Law of Gravity, $F = g\,M\,m/r^2$, where M is the mass of the Earth, one observes that the mass of the falling object m cancels out of both sides and the acceleration a of the mass m is proportional to quantities not relating to m at all ($a = g\,M/r^2$); hence, proving Galileo's Law of Falling Bodies.

Francis Bacon and Rene Descartes

During the same time of the previous discoveries and voyages, numerous philosophers were promoting the Scientific Method to understand nature. Two men stood
out, Francis Bacon (1561–1626) and Rene Descartes (1596–1650).

Bacon was an English politician, philosopher and promoter of empirical science.
He proposed that knowledge could best be advanced by experiment, as opposed
to, interpreting and refining knowledge from established authorities. His answers to
the question of how one makes true observation and conclusions about nature were
published in his *Novum Organum* in 1620.

Descartes was a French mathematician who invented the Cartesian coordinate
system that enabled mathematical functions to be expressed as algebraic equations.
In his *Discourse on Method* (1637), Descartes pointed out—only accept anything
being true until proven.

The result of this thinking was the beginning of an intellectual change in society,
produced by clarity of reason and the relegation of faith to a subservient position.
The Scientific Method provided a mechanism for finding the truth, independent of
the teachings of the ancients and the church.

Royal Society of London

Bacon's proposal that knowledge could be advanced by experiment motivated the
establishment of the Royal Society by Robert Boyle (1627–1691), Christopher Wren
(1632–1723) and William Petty (1623–1687) to improve knowledge for the benefit
of humanity. These English scientists were followers of the Baconian method of
experimental learning.

The courtier and scientist Sir Robert Moray (1608–1673), who resided at White-
hall, reported that King Charles II (1630–1685) was informed about the project and
approved of it. In July 1662, the Royal Society's charter was formally authorized by
Charles II. Members were to be referred to as Fellows of the Royal Society. Now
enshrined in law, the Society was permitted to authorize individual publications—a
major right in an age when all forms of communication were state regulated. A second
charter, which the King authorized in April 1663, emphasized that the Society, now
called this Royal Society of London, would have close links to the Crown.

The men of the Society were motivated by their belief that the increase in natural
knowledge would automatically lead to improvements in manufacturing, commerce
and trade. The Society became a forum for sharing discoveries and ideas and a
center for coordinating national and international research. At a local level, actual
experiments were performed at weekly meetings.

In March 1665, the *Philosophical Transactions* was published by the Society, becoming the first scientific journal in Europe and eventually the longest running one. The journal increased the Society's reputation and its role as a hub of scientific communication. It was useful in settling arguments over the priority of a discovery by pointing to a publication in one of its journals. It also provided an important platform for an international culture of scientific knowledge and a public forum for the dissemination of ideas. Its publication was the beginning of uncensored collective learning.

The English were not singular in their support of science. A few years after the Royal Society was founded, the Academie of Sciences in Paris was established in 1666 by Jean Baptiste Colbert (1619–1683), the financial controller of Louis XIV's (1638–1715) regime. With Louis's patronage, it formalized under government control previous private gatherings on scientific issues. In 1699, the Academie received a constitution in which six subjects were supported: mathematics, astronomy, mechanics, chemistry, botany and anatomy. The Academie was committed to advising the French government on scientific issues and advancing matters of science.

The Enlightenment

The voyages of discovery and the Scientific Revolution generated a new way of thinking about the world. This expanded view helped usher in the Age of Reason or what is referred to as the Age of Enlightenment. John Locke, Baron de Montesquieu, Voltaire and Denis Diderot were four philosophers who significantly influenced the period through their writing.

John Locke

John Locke (1632–1704) was an English physician and political theorist who used empiricism, common sense and observable experience, that was verifiable, to discern knowledge. In his *Essay on Human Understanding*, he proposed that differences among men were due to differences in education and experience rather than to birth advantages. Although completed in 1666, its publication was delayed until 1690. In his *Second Treatise on Government* (1690), Locke argued that governments were made to insure man's natural rights, namely rights to life, liberty and property. These inalienable rights became a principle of Liberalism.

Baron de Montesquieu

Baron de Montesquieu (1689–1755) was a French political philosopher, also grounded in Liberalism, who wrote that the primary purpose of government was to support law and order, liberty and property. His political ideas were progressive and radical for the times. He conceived the idea of the separation of political powers within government into executive, legislative and judicial. He is best known for *The Spirit of Laws* published in 1748. He reasoned that by keeping the powers separate, each branch would provide a check on the other, thus providing a political mechanism for enabling Enlightenment ideals.

He believed that combing these three governmental powers would lead to authoritarianism, as in the French monarchy of Louis XIV. In contrast, Montesquieu admired the form of England's government where the king represented the monarchy, the House of Lords the aristocracy and the House of Commons that of the people. His theory on the separation of powers became the basis for shaping modern democratic governments. This concept of the separation of political powers subsequently became the cornerstone of the Constitution of the United States.

Voltaire

Voltaire (1694–1778) was a Frenchman whose voice brought the work of Locke and Newton to the French people through his writings. He thought social progress could be achieved through reason. He emphasized the importance of religious tolerance and that no authority, political or religious should be exempted from challenge by reason. These ideas were succinctly expressed in his *Letters philosophiques* published in 1734. Voltaire was a prolific writer who extolled the rights of man, material prosperity and the abolition of torture. His most famous work was *Candide* (1759), a satire and social criticism of philosophical fiction about the church, a beneficial God, clerics and war.

Denis Diderot

Denis Diderot (1713–1784) was a French writer who was among the most influential of the philosophers that questioned the religious and political authorities of eighteenth-century France. He advocated for a more compassionate political order and argued that religion founded on revelations that were supernatural were irrational and dangerous.

Diderot maintained that all citizens should have a stake in the nation and that correct information is needed to make informed decisions. He maintained that a free press and free speech were central to the dissemination of accurate information needed for rational thought and action. These beliefs were his motivation to collect all knowledge into one large publication, the *Encyclopedie*. This prodigious undertaking began to appear in a series of volumes between 1751 and 1772. Eventually 28 volumes were published, of which 17 were volumes of illustrations. He both edited and contributed over a thousand articles together with his coeditor Jean le Rond d'Alembert. The *Encyclopedie* brought the concept of science as progress to the consciousness of Western Europe.

Chapter 9
Modernity: 1700–1900 (Industrialization and the Steam Engine)

English Industrial Revolution

Preconditions

By 1700, most of the world was connected through exchange networks. The formally isolated European nations, of Spain and Portugal were interconnected with their colonies. England, France and Holland expanded through their colonies. Despite these increased associations, the world was still conventional and unchanged. Governments were traditional. Most people made their livelihood through agriculture and many were still peasants. Cities and towns contained no more than 10–20% of the population. Communication was minimally changed; the fastest way to convey a message was by a dispatched rider or boat. What transformed was only the magnitude of what was being exchanged.

This situation was about to radically change, with the development of the industrial revolution. That this transformation evolved first in one place, with England pioneering industrialization in textiles, was in contrast to the transition of the agricultural revolution 10,000 years earlier, which happened in many places at once. This change was complex and involved the coming together of many factors simultaneously.

Early commerce caught the attention of the government and bankers in England. Customs and excises taxes became the main revenues supporting the government. In Parliament, this money became the rationale for strongly supporting trade with a Navy that protected its overseas empire. By 1694, England had sufficient capital to found the Bank of England, which had the capital to provide a stable currency and ready credit for investment. Stable capital was critical to forming the seed money to assist in starting new ventures. The separation of powers in England, with its rule of law, made authorities accountable and contracts enforceable. The foundation for investment and growth was created, preventing the Crown from taking earnings.

© The Author(s), under exclusive license to Springer Nature Singapore Pte Ltd. 2024   133
T. Sanford, *A Whirlwind History of the Universe and Mankind*,
https://doi.org/10.1007/978-981-97-2674-5_9

In 1763, Britain won the Seven Years War (known in America as the French and Indian War). With this success, England replaced Spain, France and Holland to become the center of trade in the Atlantic Ocean. North America became controlled by England and this change brought extensive raw materials with reliable new markets, in particular, cheap cotton. Capital was amassed from these market investments.

Enlightenment ideals prevailed during this time, which encouraged a variety of innovations. There was no guild system to control or suppress innovative ideas of production. Financial institutions mobilized capital with the support of Parliament and encouraged commerce. Social and ideological conditions inspired development and free thinking. Numerous immigrants from the disruptions caused by the religious wars became entrepreneurs, i.e. the Huguenots, Dutch and Jews. Many believed God was revealed through the book of nature. Science was viewed as important. Early on, the Crown honored men of learning i.e. Francis Bacon and Isaac Newton, who were knighted for their contributions in 1603 and 1705, respectively. Prizes were offered and laws passed like the Longitude Act of 1714, which determined the location of a ships longitude.

A necessary ingredient for industrialization was labor. Serfdom was abolished by Queen Elisabeth (1533–1603) in 1574, which freed the peasants. Initially, the manors consisted of largely self-sufficient farms. Peasants paid rent or labor services to the manor lord in exchange for farming the land. Since the sixteenth century, a legal process of consolidating small landholdings into larger ones by the formal process of Enclosure had been underway. Peasants on many small farms used land held in common to survive. As wealthy landowners bought more land, including that of the commons, the peasants were driven from the land to become wage earners on larger estates or in towns.

On the larger estates, the landowners were generally less interested in producing food for their own use. Instead, crops were grown for a market. Efficiency in production became essential in reducing costs. Seeds were planted in rows with machines drawn by horses. Crop rotation and irrigation were implemented. Better drainage and fertilization improved the soil and thus output. Agricultural production expanded, increasing by a factor of 3.5 from 1700 to 1850. With increased efficiency in farming, the number of people needed in agriculture dropped from 61 to 29%. The cost of food decreased and the peasants evolved into potential industrial laborers and buyers. With improved nutrition, fewer pandemics and no major wars fought on English soil, Britain's population doubled from 1750 to 1800.

Textiles

With these innovations, the textile industry became the first to mechanize. Hand spinning and weaving were replaced by newly invented machines. Until 1733, weavers moved the shuttle back and forth by hand. That year, the English machinist and engineer John Kay (1704–1779), invented the flying shuttle, which doubled the speed of weaving. The following year, the English weaver, carpenter and inventor, James Hargreaves (1720–1778), developed the spinning Jenny. The Jenny used a large wheel to spin many spindles of threads at the same time, increasing the speed of textile manufacturing, primarily for cotton. In 1769, the English entrepreneur and inventor, Richard Arkwright (1732–1792), patented the water frame, which was capable of producing stronger yarn. It represented an improvement over the spinning Jenny, which produced weaker thread. The water frame was powered by a waterwheel that spun a shaft. As these processes for spinning yarn were mechanized, the need for human labor was reduced. Arkwright became wealthy and in 1786 was knighted for his achievements that helped propel England into the industrial revolution.

After visiting Arkwright's spinning machines, the English inventor Edmund Cartwright (1743–1823) thought he could develop a similar machine for the weaving process. In 1785, he patented a power loom that mechanized the weaving function. In 1809, Parliament awarded Cartwright 10,000 pounds for his contributions to the British textile industry. He was elected a Fellow of the Royal Society in 1821.

These process accelerated the production of textile manufacturing and had a large impact in the nineteenth century. In 1803, there were only 2300 power looms in Britain, however, by 1833 there were 100,000 being used in British textile factories. The textile industry, benefited immensely, as it could now generate the faster mass-produced clothing in factories.

Only a short period of time elapsed before James Watt's (1736–1819) newly invented steam engine would have enough power to be attached to both the spinning (water frame) and weaving (power loom) machines. This last step occurred during the late 1790s and the early 1800s. Between 1780 and 1800, industrialization reduced the price of textiles made with cotton by 80%. By 1850, the British were importing ten times the cotton as 50 years earlier, thus began the industrial revolution.

The Steam Engine

During the Little Ice Age of 1350, temperatures cooled in many parts of the world, including Europe. Throughout the cooling, the British burned wood for heating. Contributing to the use of wood was ship construction. Charcoal made from wood was used in the smelting of iron ore to eliminate its impurities. By 1800, only 5–10% of Britain remained covered in forests. To save the wood for lumber and charcoal, the British began using its vast resources of easily accessible coal for heating.

Abraham Darby (1678–1717) learned how to smelt iron by burning coke, which was made by heating coal in the absence of air. The use of charcoal was replaced and an additional use for coal was generated. Eventually, the mined surface veins of coal were depleted and deeper shafts followed the veins until groundwater slowed operation. Initially, inefficient steam pumps, using a fraction of the mined coal, were utilized to extract the water.

With the need for an effective method to mine coal, the Scottish instrument maker, James Watt, from the University of Glasgow, developed the first efficient steam engine in 1776. His engine was applied to a multitude of industrial processes. Factories, which previously had to be near flowing water, could now be built anywhere. Requests for his engine came not only from the textile industry but from iron mills, paper mills, flower mills, waterworks, canals and even distilleries. By 1800, 500 of the new engines were being used in England and, by 1830, the main source of British power was the steam engine. Between 1800 and 1900, engineers and entrepreneurs developed steam engines that became ten times more efficient and weighed a fifth as much per power output.

Improved steam engines enabled entry into deeper mines, which lowered the cost of mining coal. The quantity of coal mined increased by a factor of 55 between 1800 and 1900. The efficient use of coal by the steam engine accelerated the nascent industrial revolution. A factory system developed, in which workers labored under one roof with the steam engines powering many of the machines.

The steam engine resulted in new breakthrough technologies such as the steam locomotive and the steamship. Rails with cars, first developed to bring coal to the surface from mine pits, were now coupled with steam locomotives for intercity transport, during the 1820s. By 1830, the first commercial rail line connected Manchester with Liverpool. By 1850, railroads were linking cities through out Britain (e.g., Birmingham, Bristol, Glasgow, Liverpool, London, Manchester and Sheffield). By 1870, approximately 21,000 km of track had been laid. Trains provided inexpensive transportation to carry raw materials, finished manufactured goods and people. Industrialization was further incentivized.

In the early 1800s, steamboats were developed, as the steam engine turned the paddle wheels. Progress was swift. By 1840, the joint British and North Royal Mail Steam Packet Company launched the RMS Britannia, a wooden hull, coal-powered side paddle wheeler. It provided regular passenger and cargo service between Liverpool, Halifax and Boston, crossing the Atlantic Ocean in under 10 days. The average speed was 20 km/hr, which set a world record. Seven years later in 1847, the SS Great Britain became the first screw-driven, iron hull steamship to cross the Atlantic with speeds of 19–20 km/hr. These steamships, with their continued evolution and decreased dependence on wind patterns, opened new trade routes and became catalysts for globalizing trade after 1870.

In summary, between 1744 and 1860, the population of England tripled from 6 to 20 million people. During approximately the same period, the income per person doubled, increasing the buying power of individuals. In 1700, less than 17% of the population lived in towns. By 1850, the majority either lived in towns or in cities, and by 1900, one in five Britains resided in London, bringing its population to 6.5 million. London became the largest city on Earth. Britain became the prototype for the rest of the world to follow. With this new industrial strength, it led the nineteenth century in becoming the greatest non-contiguous empire in the history of the world.

This essential transformation was based on the efficient conversion of stored energy in coal to raw power via the coal burning steam engine that was first developed in 1776. A primitive steam locomotive could produce approximately two hundred times as much power as that generated by a plow team of two horses.

The political and social conditions that fostered encouragement to scientists, engineers and entrepreneurs, with their questioning of authority and attitude of experimentation, was essential to achieve these changes. Although the preconditions had been evolving slowly, once the steam engine was developed, the transformation to industrialization took only a hundred years, from approximately 1780 to 1880.

Birth of a New Society

As farming mechanization advanced, its displaced workers moved to towns where factories were being located. Handwork at home in cottage industries shifted to the factories. The working conditions of the factories became regimented and disciplined, with a clock defining the hours of work instead of the Sun. Industrialization became competitive, resulting in managers needing to keep labor costs low by requiring long work hours at low wages, with few safety features.

This evolving industrialization produced new wealth and created new classes of society: a working class, a middle class and a wealthy class who wanted a greater degree of social influence. The managers of businesses, small and large, became the middle class. A bureaucracy composed of professionals, engineers, scientists and inventors was created. In general, these entrepreneurs worked for pay, were educated, advanced based on merit and owned property. They supported political reforms, including rights to vote and often served in Parliament. They saw that progress and modernization were based on liberty and social harmony, with a responsibility between the individual and state.

The towns were not prepared for the masses of people migrating into their interiors, which were being transformed into cities. The living conditions became slums with unhealthy environments. Initially, workers lived in poverty as the growing middle class became managers and wealthy. As the laborers began to demand reforms, a mechanism was needed to resist the excessive restraints on wages and poor working conditions. Various trade unions were formed to accomplish this. Those changes occurred slowly but, by 1867, the legal status of trade unions was established and evolved to benefit both labor and management. As a result, the working classes began to experience the benefits of the industrial revolution.

With time, industrialization had a positive effect at all levels of society. The production of goods expanded, generating a consumer society that had amplifying wealth. The standard of living increased, with access to healthier foods, improved housing and cheaper goods. Better education and health care augmented the life span. The middle and wealthier classes benefited immediately, while that of the laborers improved more slowly.

In the late eighteenth century with a growing population and the influence of industrialization, the British standard of living increased and evolved into an era of growth in per capita income. With the increase in sustained growth and increasing income, industrialization was quick to spread outside England, first to Western Europe and the United States, then to the rest of the world.

The English model became a template for the world to follow, creating new wealth, goods and services, new cities and new social structures. England broke the Malthusian cycle that occurs when population expansion surpasses agricultural production, causing population to be restricted by famine. The industrial revolution, initiated an associated socioeconomic order that largely forms the civilization known today. This evolution is as significant as the earlier developments of fire and agriculture.

Industrialization Spreads

Western Europe

The countries of Europe were often in competition with one another. The benefits of industrialization were quickly observed and spread, first to England's neighbors and colonies. Belgium had both coal and iron deposits, a history of textile production and was able to rapidly industrialize. France was hampered by its lack of coal and iron, an inefficient agricultural system and the French Revolution. Investments in industrial innovations were restrained due to uncertainties in its political state. The ideals of the Enlightenment, the stimulus for the French Revolution, however, were beneficial. Industrialization in France began in 1810 in the textile industry; by 1848, France had become an industrial power.

Industrialization in Germany commenced in 1840, when serfdom was abolished. With its rich reserves of coal and iron, Germany began building railroads. Once Germany achieved national unity in 1871, the industrial output expanded rapidly. By 1900, it was outproducing Britain in steel and became a leader in the chemical industry.

Italy, especially Florence and the maritime city-states like Venice, had flourished with respect to other parts of Europe until 1600. After 1600, Italy began to stagnate and, by the nineteenth century, the economy was less advanced than the rest of Western Europe. The lack of significant coal and iron deposits, a largely illiterate population and an extended history of political fragmentation were significant impediments to industrialization. A sporadic industrial development occurred in Naples in 1850 and in the Milan-Turin-Genoa triangle prior to national unification in 1861. However, Italy did not began to industrialize until 1890, and then only in the northern half. The feudal land system that ruled in southern Italy from the Middle Ages deteriorated after unification and has yet to be resolved.

United States

After independence, the U.S. followed a similar trajectory as England, starting with the textile industry in New England. The rivers of the northeast provided the water power needed for the textile mills. Samuel Slater (1768–1835), who learned about cotton mills while working in English mills as a youth, established, along with others, the first water-powered cotton-spinning mill in 1790 at Pawtucket, Rhode Island. Slater's knowledge of production and management led to the manufacturing of machinery that was sold to other cotton mills, setting the foundation for New England's textile industry.

The United States was rich in natural resources, especially coal and iron, leading to railroad construction beginning in 1830. The government initiated land grants to start the process of laying the rail lines. Private companies provided the funds and loans arrived from European bankers. During the Civil War (1862–1865), the Pacific Railroad Act of 1862 provided grants of land and loans to build a transcontinental line. Construction began in 1866, after the war ended. The Union Pacific rail line started in Omaha, Nebraska and the Central Pacific started in Sacramento, California. The two lines met in Promontory, Utah in May 1869. Travel by railroad from the Atlantic to the Pacific coasts became possible for the first time.

Industrialization, especially during the Civil War, accelerated the production of munitions in the northeast. After the war, the arms producers looked elsewhere for buyers. With minimal government regulation, industrialization was greatly increased. By the end of the nineteenth century, the United States surpassed Great Britain's output of manufactured goods, generating nearly 24% of that used by the rest of the world, in contrast to Britain's 19%.

Empires Late to Industrialize

Russia

Prior to 1860, the Russian people were predominately serfs bound to the lands of their masters. The aristocrats imported manufactured products and machinery, which they bought in exchange for grain and timber. After losing the Crimean War (1853–1856), Tsar Alexander II (1818–1881) freed 47 million serfs and introduced reforms that promoted industrialization by 1866.

Industrialization began by: generating banks, hiring foreigner engineers and building railroads. In 1891, during the reign of Alexander III (1845–1894), the world's longest railway linking Moscow to Vladivostok through Novosibirsk was started and completed in July 1904. By 1900, Russia's share of world's manufacturing output reached 8.9%, ranking fourth in the world above that of France. Using commercial trains and steamships, it was now possible to circle the globe with greater ease and speed than Phileas Fogg did in Jules Verne's novel: *Around the World in 80 days,* published in 1872.

China

China and India were two of the most advanced civilizations. By the 1400 s, they had a standard of living greater than that of Europe and by 1750 led all nations in economic output. However, by 1900, China and India were among the least industrialized and became some of the poorest regions of the world.

Specifically, by the fifteenth century, the Ming Dynasty (1368–1644) of China had the world's largest economy. Their ships dwarfed the early Portuguese and Spanish caravels in size. In the early fifteenth century, China launched a number of expeditions to the west, establishing diplomatic relations with southeast Asia, India, the Middle East and the east coast of Africa. With a change of emperors in 1433, these forays to the west ended and resources were concentrated on defending its northern borders against nomadic warriors.

There was little incentive to venture beyond the Indian Ocean in a midst of the riches within China. China became inward looking and insular, seeing little to be gained by more exploration beyond its borders. China reigned supreme in east Asia, with minimal competition for its produce. With a growth in population that exceeded England's, after 1300 there was little motivation to develop labor-saving devices that could increase labor productivity. Trade with Europeans was restricted to a limited number of port cities. The elite class in China had a general resistance to industrialization, therefore economic development in China did not begin at a significant level until the 1950s, when approximately 83% of the workforce was still employed in agriculture. By 2018, however, that percentage dropped to 26% and China once again, became a major world power. In 2022, China is the world's second-largest economy, competing with the United States for first place in global GDP (Gross Domestic Product).

India

By the late seventeenth century, the majority of the Indian subcontinent was reunited under the Mughal Empire (1526–1720). By 1700, India evolved to become a large economy and manufacturing power, producing 25% of the world's output.

During the 1600 s, Britain traded extensively with India. After the Battle of Plassey in 1757, the British East India Company (which was operating as a military authority in parts of India) achieved a significant victory over the Bengals and their French allies. By making military, trading alliances and treaties with the numerous independent Indian states, Britain began almost two centuries of gradual increasing British rule. After suppressing the Indian Rebellion of 1857, the East India Company was abolished. Direct rule of India by the British government was established, with India becoming the Jewel in the British Crown.

During the nineteenth century, India experienced de-industrialization and loss of various craft industries, due to competition with British manufactured products.

During British rule, however, a vast railway structure was supported that represented the limited industrialization undertaken.

In 1947, India gained independence from Britain. Initially, the new Republic of India used central planning as the guide for development. After the economic crisis of 1991, the government initiated economic liberalization, which allowed India to emerge as one of the fastest growing industrial powers. In 2022, although approximately 60% of the population still make their livelihood from agriculture, India accounts for a fifth of the the global GDP.

Ottoman

The Ottoman Empire, after the conquest of Constantinople in 1453, proceeded to control the lands of southeast Europe (the Balkans) and western Asia. This expansion, included the ancient lands of Mesopotamia, the Levant and the lands surrounding the southern Mediterranean Sea, and the coast from Egypt to Algeria (until the early twentieth century).

At its peak in the sixteenth century, the Ottoman Empire became one of the largest economic and military powers in the world. They gave rise to the "feared" Turks by the Europeans, during the time of Columbus.

Their economy was primarily agrarian, based on surpluses produced by the individual lands they controlled. The Empire, centered in Constantinople, was governed by a bureaucracy of officials responsible to the sultan (emperor). Internally, the society was thought to be superior to anything the outside world could produce. The Islamic elite class had all the services, goods and products they desired. Inventors and entrepreneurs, who were generally employed by the wealthy, had little motivation to innovate or introduce new ideas or technology. The majority of the Ottomans felt little need for the Empire to change.

When the printing press was developed in the West, Muslim legal scholars and manuscript scribes were opposed to its use, because Allah had said it all. The flow of information was thus curtailed. In 1515, the sultan was persuaded to issue a decree that imposed the death penalty on anyone using a printing press to publish books in Arabic or Turkish. As a result, Constantinople did not get its first printing press until 1726.

With time, the ruling class became isolated from outside developments. They became blind to the advances in technological, scientific, commercial or industrial developments that occurred after the Reformation. On the battlefield, they did not connect the failures experienced with the growing superiority of the Western armies. Once enlightened in the case of fighting, however, the Ottoman Empire adopted specific European weapons and techniques of undeniable advantage.

After the Europeans discovered ocean routes to India, China and the Spice Islands, the Ottoman Empire began to slip into a long, slow decline. During the decline, the Greeks were the first to rebel in 1821. As the Industrial Revolution was advancing throughout Europe, the economy of the Ottomans continued fixated on agriculture.

During World War I (1914–1918), the Ottomans joined the Central Powers, Germany and Austria-Hungary. They did not have the industrial capacity in iron or steel to build the needed railways and heavy weaponry. Another limitation was a shortage of professionals, engineers and doctors. Only 5–10% of the population were literate at the time the war began. Moreover, the Empire joined the losing side. So when the war ended, their vast territories were partitioned among the British, French, Greeks and Russians.

The Turkish War of Independence (1919–1923) followed among the Turks and occupying powers. In 1922, after 600 years of Ottoman rule, the last Ottoman sultan was deposed. The following year, the modern secular state, the Republic of Turkey emerged, with Mustafa Kemal Ataturk (1881–1938) as its first president. The capital was set in Ankara, Turkey. The size of the territorial was reduced to that of Anatolia (modern day Turkey) the homeland of the majority of the Turks. Ataturk was progressive and introduced reforms that both westernized and secularized the country, including a new writing system for Turkish based on the Latin alphabet, giving full rights to women and symbolically abolishing the Ottoman fez.

Two Revolutions and Two Canals

Several events occurred, during the early Industrial Revolution, that significantly enhanced the present social and physical richness of the world. The American Revolution of 1775–1783 and the French Revolution of 1789–1799 helped to continue the promotion of the ideals of the period of the Enlightenment. The construction of the Suez Canal in (1859–1869) and the Panama Canal in (1888–1914) illustrated the newly found power of the Industrial Revolution to significantly modify the physical world for the benefit of all humankind.

American and French Revolutions

The American Revolution furthered the liberal ideals of democratic governance, which included separation of church and state and support for the concept that all people have the fundamental rights of freedom of religion, speech, assembly and equality under the law. The 1776 American Declaration of Independence solidified these ideals with the words that all people have the unalienable rights of "Life, Liberty, and the Pursuit of Happiness."

The French Revolution was influenced by the American Revolution. France reinforced many of those ideals and aspirations by promoting the concepts in the motto "Liberty, Equality, Fraternity" and in the Declaration of the Rights of Man and of the Citizen. Both countries saw themselves as initiators of liberty by promoting republican ideals. That connection was exemplified by the French gift of the Statue of Liberty to the American people in 1881.

The changes in France, following their revolution, were extensive. The feudal system was terminated. The privileges of noble birth were abolished. The form of government evolved from a monarchy to a republic. The church lost its monasteries and farmlands and the bishops were elected by the people. Repercussions occurred throughout Europe. Awareness of French liberal ideals about legal equality, objections to oppressive governments and the benefits of having a parliament with a constitution spread. These ideals furthered the replacement of absolute monarchies with republics and liberal democracies.

The French Revolution promoted a uniform system of measures and weights. Prior to this time, a decimal-based system of units originating from the Romans was chaotic, not only in France but throughout Europe. During the early stages of the revolution, the right of the aristocrats to control the units used in their domains was abolished. By 1795, the unit of length would be defined as the meter (from the Greek word *metro* for measure), the liter for volume and the gram for mass. The meter would be defined as the equivalent of one ten millionth the distance from the North Pole to the equator and so forth. On December 10, 1799, the French adopted this metric system of units as part of their endeavor towards modernization.

Suez and Panama Canals

The idea for a canal linking the Mediterranean Sea to the Red Sea originated in the era of the Egyptian Pharaohs. After development of the steam engine and the associated industrial organization, constructing a canal became more practical. With growing commerce between Europe and Asia and expansion into Africa the motivation to make that connection increased as it would shorten the journey around the African continent.

In 1854, Ferdinand de Lesseps (1805–1894), a French diplomate and engineer, was granted an agreement from the Ottoman Governor of Egypt to construct a canal across the Isthmus of Suez. Two years later, an international consortium, the Suez Canal Company, was formed and a group of engineers drew up plans to construct a 193 km canal between Suez and the Red Sea. In 1859, the digging began, using forced labor with picks and shovels. In 1863, that labor was replaced with European workers, using custom-made coal and steam-powered shovels and dredgers. With the added mechanical power, the project was completed in 1869. Over the duration of the project, more than 1.5 million people were involved, averaging 30,000 working at any one time.

After completion, the Suez Canal became the world's most heavily used sea-lane route, greatly shorting the shipping time between the north Atlantic and Indian Oceans for the benefit of all. In 1875, Britain bought Egypt's interests in the Suez Canal and together with France, continued to control the canal until 1956.

The idea for creating the Panama Canal, a water passage between the Atlantic and Pacific Oceans, dated to shortly after the Spanish first discovered the Pacific in 1500. Although shortening a shipping route between Europe and Asia remained a worthy goal, crossing the jungle and mountains of Central America was impractical. But, with the success of the Suez Canal, the French, again under the leadership of Ferdinand de Lesseps, attempted to build a sea-level canal in Panama in 1881. The combination of yellow fever, malaria and other tropical diseases contributed to the loss of 20,000 lives, which ended the French attempt eight years later in 1889.

In spite of these setbacks, interest in a canal continued. Both the United States and Britain business exerted sustained political pressure for such a canal to ship goods cheaply and quickly between the two oceans. The President of the United States from 1901 to 1909, Theodore Roosevelt (1858–1919), was able to achieve that goal. The Treaty of 1903 between Panama and the United States granted America a 16 km wide strip of land for the canal. Work began on the 64 km long canal in 1904, which was completed ten years later in 1914.

The success of the Panama Canal was due to a number of discoveries and innovations. It was the recognition that mosquitoes carried the deadly diseases, which were eradicated by fumigating homes, cleaning and eliminating pools of water. The newly discovered dynamite, together with massive steam shovels, were used to blowup and excavate the mountainous regions along its length. Almost 184 million cubic meters of rock and dirt were removed with 27 tons of dynamite. Altogether, approximately 3.4 million cubic meters of concrete were poured to construct the locks.

The Panama Canal was an engineering marvel with a unique design, that used artificial lakes to minimize excavations and locks that operate by gravity flow from lakes fed by river water. On the Atlantic side at Colon, a series of three locks raised vessels 26 m to Gatun Lake, which was formed by a dam, making it the largest artificial lake in the world. On the Pacific side, the locks lowered vessels 9 m to Miraflories Lake, which was formed by a dam at an elevation of 16 m above sea level. Vessels then passed through a second set of locks before being lowered to sea level near Panama City. Between the two sets of locks, near the Pacific side, the canal traversed the Gamboa Cut at the Continental Divide. The divide, originally 110 m high, was reduced to 13 m at the location of the canal.

Between 1904 and 1913, 56,000 workers were employed. Over that period, diseases and accidents claimed the lives of more than 5600. In 1914, the canal opened to traffic, more than 30 years after the first attempt made by the French. At the time, the Panama Canal was the greatest engineering project humankind had accomplished. It's completion illustrated the growing American economic power and technological progress.

In 1977, the United States and Panama signed a new treaty that abolished all former treaties. It recognized Panama as the territorial sovereign of the canal zone, granting the United States the right to continue operating, maintaining and managing the canal. The new treaty provided a vital component in the growing trade routes of the twentieth century, ensuring freedom of passage to all.

In building the canal, the replacement of gunpowder with the more powerful dynamite was an important factor. Using dynamite reduced the chance of an accidental explosion and was cheaper to manufacture. Dynamite became the product of choice in mining, quarrying and construction. Its creator, the Swedish chemist Alfred Nobel (1833–1896), became very wealthy. When Nobel died he bequeathed a large fraction of his fortune to establish the Nobel Prizes in 1895 in recognition of the significant advances in the fields of physics, chemistry, medicine, literature and peace. Today, the Nobel Prizes are regarded as the most prestigious awards in their respective disciplines. With the development of science, these prizes provide a useful guide to their evolving discoveries.

Following the pioneering work of prodigious scientists, like Galileo and Newton, physics continued developing throughout the eighteenth century. During the Enlightenment age, scientists, engineers and entrepreneurs were encouraged to explore and discover new possibilities. The industrial revolution in the nineteenth century created the engineering knowledge and generated the instruments to probe nature in different and fundamental ways. As a result, the new aspects of physics and the dependent sciences of chemistry, biology and medicine advanced precipitously. In this progression, physics paved the way.

Part III
Physics

Chapter 10
Physics: 1700–1900

Thermodynamics

Energy

With the development of the steam engine, the industrial revolution made use of important concepts of energy and power, even though these were not yet fully understood or defined. There was strong motivation to advance the efficiency of these early steam engines. Understanding the conservation of mechanical and heat energy, during their development, gained meaning and importance. During this time the field of thermodynamics evolved.

In the seventeenth century, the building blocks for understanding energy were already being explored. The German mathematician Gottfried Leibniz (1646–1716), discovered in many mechanical systems (of several masses m_i each with different velocities v_i) the quantity that sums the products of m_i times v_i^2 was conserved. Today, the sum is recognized as related to the total kinetic energy of a system. Independent of Newton, Leibniz also developed the differential and integral calculus. The symbols for these mathematical operations, used today, were those he designed.

The quantity $mv^2/2$, applying the calculus, is easily derived from the definition of energy, which became the ability to do work W against a force F over a distance x. In one dimension, the work done on an object of mass m against a force F over a fixed distance x is defined mathematically by the integral: $W = \int F \, dx$. Using Newtons second law of motion $F = ma$, where a is the acceleration (dv/dt) of a mass m and changing variables in the integrand dx to dv, the integral is solved as $W = m \, v^2/2$. This quantity is the kinetic energy a mass possesses due to its motion with velocity v; or energy is the work needed to accelerate a body of mass m from rest to velocity v. In 1802, the term energy was first used by the polymath Thomas Young (1773–1829) in lectures to the Royal Society. Complementing energy, the time rate of change of performing work or in providing energy is power.

© The Author(s), under exclusive license to Springer Nature Singapore Pte Ltd. 2024   149
T. Sanford, *A Whirlwind History of the Universe and Mankind*,
https://doi.org/10.1007/978-981-97-2674-5_10

The SI (International System) of Units known as the metric system, is the international standard for measurement. Its unit of energy is the Joule (J), the unit of power is the Watt (W) and the unit of force is a Newton (N). One Newton is the force that gives one kilogram an acceleration of one meter per second squared; one Joule of energy is the work done by a force of one Newton acting through a distance of one meter; and one Watt of power is defined as generating one Joule of energy per second. These names were given to honor the men who gave significants to them.

Heat

Prior to the early nineteenth century, heat was believed to be a substance identified as the caloric, which was that transferred from a hot object to a cooler one. The American-born British physicist Sir Benjamin Thompson (1753–1814) demonstrated that vast quantities of the caloric could be produced in frictional processes. This observation did not agree with the caloric being conserved. Eventually, the English physicist James Joule (1818–1889), using energy from falling weights in precise mechanical experiments to heat fluids, proved unequivocally that kinetic energy was what was being converted to heat energy. Today, the word heat refers only to energy in transit. Once heat is transferred to an object, it becomes the internal energy of the object that is increased. As the object becomes hotter, its temperature rises. By the mid nineteenth century, one calorie (cal) was defined as the amount of heat required to increase one gram of water one degree Celsius. The experiments of Joule established that this temperature increase was due to imparting 4.17 J of energy to the water.

In 1841, Joule established the First Law of Thermodynamics, which states "The change in the internal energy of a system is equal to the heat added to the system minus the work done by the system." The total energy of a system is conserved. Later, the great British physicist, William Thomson (1824–1907), concurred with Joule's First Law. In 1851, William Thomson articulated the Second Law of Thermodynamics: "...heat does not spontaneously flow from a colder body to a hotter one."

There is a tendency in an isolated system to degenerate into a more disordered state. The German physicist, Rudolf Clausius (1822–1888), developed the concept of Entropy as a measure of the disorder or randomness of a system. The manifestation of entropy's increase in macroscopic phenomena is best illustrated by nature's continuance of time, resulting in change. Following William Thomson, the Third Law of Thermodynamics states "...a state of total order or minimum entropy defines an absolute zero of temperature; it is approximately minus 273 degrees Celsius."

Atoms, Elements and Molecules

The idea that matter was atomic; that it was made up of simple, indivisible, indestructible dates to the time of Democritus (460–370 BCE) in ancient Greece and Rome as expressed by Lucretius (94–51 BCE) in his epic *On the Nature of the Universe*. This idea was partially accepted by Galileo and Newton but it was not until French chemist Antoine Lavoisier (1743–1794) and English chemist and physicist John Dalton (1766–1844) that the atom was given its modern form.

Earlier, the English chemist Joseph Priestley (1733–1804) was inventing and exploring various gases and mixtures, including soda water (water with carbon dioxide). In a series of experiments ending in 1774, he discovered that air was a combination of gases, one of which he identified as oxygen. Priestley established its early connection to combustion and respiration, showing that plants could absorb carbon dioxide and release oxygen.

Lavoisier repeated Priestley's experiments and designated oxygen as its name in 1778. He was noted for exploring the role oxygen plays in combustion. He could not find a way to reduce oxygen and hypothesized that oxygen must be an element—a singular substance that could not be further decomposed by chemical changes. He discovered other elements including hydrogen, nitrogen, carbon and silicon. In his book, *Elements of Chemistry* (1787), he compiled the first list of all elements previously discovered. He assisted in revising and standardizing the chemical nomenclature and promoted the metric system of measurement, during the early French Revolution. Today, Lavoisier is viewed as the father of chemistry.

Building upon the work of Lavoisier, Dalton developed the atomic theory of chemical elements. His work with gases motivated him to assert that every form of matter (whether gas, liquid or solid) was made of small individual particles and that each element was made of unique atoms with features specific to that element. To support this concept, Dalton introduced the idea that atoms of different elements could be distinguished by their varying atomic weights A, which was published in his *New System of Chemical Philosophy* in 1808. The atomic weight (also called atomic mass) is the average mass of a chemical element's atoms relative to a standard. Since 1961, the standard unit of atomic weight has been one-twelfth the mass of an atom of the isotope carbon-12. (Carbon-12 is composed of 6 protons, 6 neutrons and 6 electrons.) Dalton theory further stated that elements can combine into molecules having fixed atomic weights that are simple ratios of whole numbers. The molecule was referred to as a group of atoms bound together, representing the smallest basic unit of a chemical compound that takes part in a chemical reaction.

In 1811, the Italian scientist Amedeo Avogadro (1776–1856) proposed that regardless of the nature of the gas, a volume of gas (at a given temperature and pressure) was proportional to a fixed number of atoms or molecules. This number became known as Avogadro's number and was defined as the number of molecules in one mole (defined as its molecular weight in grams) of any substance. Alternatively, one gram of hydrogen gas contained Avogadro's number of molecules. This number was measured in 1865 to be equal to 6.02×10^{23}.

In 1869, the Russian chemist Dmitri Mendeleev (1834–1907) presented a paper to the Russian Chemical Society titled *The Dependence between the Properties of the Atomic Weights of the Elements*, which described the known elements according to their atomic weight. In this format, the elements exhibited many periodic properties. Elements that have similar chemical properties either have their weights increasing regularly or have similar atomic weights. The magnitude of the atomic weight controls the character of the element. Based on gaps in his table he was able to predict properties of those elements that were anticipated to fill those spaces. Gallium and germanium were discovered in 1875 and 1886, respectively, in agreement with those expectations. Mendeleev was considered to be the father of the modern day Periodic Table.

Temperature and the Ideal Gas Law

The Ideal Gas Law related the behavior of gas in a given volume to its temperature and pressure, tying together the aforementioned concept of atoms. This knowledge was a guide to improving the efficiency of steam engines and provided a foundation for developing instruments like barometers. This law grew out of numerous experiments. In 1660, the Anglo-Irish scientist Robert Boyle (1627–1691) demonstrated that in a closed system of gas, when compressing a gas while keeping the temperature constant, the pressure P increases inversely with volume V: $P \sim 1/V$. In 1802, the French chemist and physicist Gay-Lussac (1778–1850) determined that if the mass m and volume V of a gas were held constant, then gas pressure P increases linearly with temperature T: $P \sim T$. Combining Boyle's and Gay-Lussac's Gas Laws, the Ideal Gas Law was formulated as $PV = NkT$ in 1834. In the equation, N was the actual number of gas molecules (which was closely related to Avogadro's number) and k was a constant, later called the Boltzmann Constant, measured experimentally to have a value of 1.38×10^{-23}. This law was remarkably simple in form and value of k, both being nearly identical for all gases.

A decade later, Ludwig Boltzmann (1844–1906) was born. He was an Austrian physicist who became instrumental in formulating atomic descriptions of macroscopic thermal phenomena. He helped to initiate the field of Statistical Mechanics in 1870, which described the micro-behavior of thermodynamics. The majority of his work was collectively published in his *Lectures on Gas Theory* in 1896.

Following Boltzmann, the Ideal Gas Law is explained in microscopic terms, using Newton's Laws of Motion and making the following assumptions: the particles of mass m in a container of volume V with pressure P are structureless, they do not interact with each other, they move in random directions with average velocity v, and they make elastic collisions with the walls of the container, conserving momentum and energy. Under these conditions, Boltzmann determined $PV = N\,2/3\,(m\,v^2/2)$. Combining this relation with the above empirical relation $PV = NkT$, the above Ideal Gas Law, becomes: $m\,v^2/2 = 3/2\,k\,T$. This rewritten equation indicates that temperature is fundamentally a measure of the average kinetic-energy associated with the translational motion of the molecules in the container. It illustrates why gas obeys such a simple law. This explanation was a significant step beyond the world of Newton. His theories took no account of temperature, nor did it give any explanation of the connection between the pressure of a gas and the kinetic energy of gas molecules.

Light

In 1655, the Dutch astronomer, mathematician and physicist, Christiaan Huygens (1629–1695), was the first to correctly describe Saturn's rings as a disk revolving around the planet. Although Galileo had noticed them in 1610, his telescope was not strong enough to detect their exact configuration. In 1678, Huygens postulated that light was a wavefront to explain the reflection and refraction of light, founding the wave theory of light. He theorized that light was composed of waves vibrating perpendicular to its direction, which became known as Huygens' Principe. In 1690, Huygens' mathematical wave theory of light was published in his book *Traite de la Lumiere*.

In contrast to Huygens, Isaac Newton in 1700, based on his optical experiments, proposed that light was made up of particles that were too small to be observed individually (Corpuscular Theory). In 1801, however, Thomas Young designed an experiment that decisively indicated that light was characteristic of a wave, supporting Huygens. He pointed a beam of light through two thin parallel slits. On a white screen, distance from the slits, alternating bands of bright and dark light bands of light were observed. Young reasoned that the alternating bands indicated that the slits were causing the the light waves to interfere with one another. Occasionally, the interference would be constructive and the light waves would add together to create a bright band. Alternatively, the interference would be destructive and the light waves would be canceled, creating dark bands on the screen. If the light were made of particles, as Newton proposed, only two bright bands would be observed.

Less known was that Young's later studies were essential to assisting the French linguist Jean-Francois Champollion (1790–1832) decipher the Rosetta stone. In 1815, Young correctly established that the demonic script consisted of "…imitations of hieroglyphics…mixed with letters of the alphabet."

Electrodynamics

The science of electricity and magnetism has its origins in the observations of Thales (624–547 BCE) of Miletus. In 500 BCE, he discovered that static electricity could be made by rubbing fur on substances such as amber. Thales was also credited with discovering loadstone's (magnetite) magnetic properties, its attraction to iron and other loadstones. Magnetite was found in Magnesia, Anatolia. The location is assumed to provide the origin of the name "magnet." At that time, no connection was made between the electric and magnetic phenomena.

Electricity

By the end of the seventeenth century, scientists had developed practical techniques of generating electricity, using friction in manual electrostatic generators. The generators operated by moving belts, to carry electric charge to a high potential electrode. The friction does not create the charge but transfers it from one object to another. These generators became instruments in the development of the new field of electricity.

In 1785, the French physicist Charles-Augustin de Coulomb (1736–1806), using a torsion balance he invented, established Coulomb's Law. The law stated that the electrical force of attraction or repulsion F between two electrical charges q_1 and q_2 was proportional to the product of the charges and inversely proportional to the square of the distance r between the charges: $F \sim q_1 \, q_2/r^2$. When the charges were both positively or both negatively charged, the force between them was repulsive; the force was attractive when carrying opposing charges. This law was similar in form to that of Newton's Universal Law of Gravity, except in the case of gravity, the force was always attractive.

Throughout the 1780s, Coulomb's results were published in a series of papers. Subsequently, the German mathematician Carl Friedrich Gauss (1777–1855) generalized Coulomb's Law to show that charged particles contained within any arbitrary closed surface generated an electric field E that exerted a force F per unit charge proportional to the charge q enclosed by the surface. The result was independent of how the charge was distributed within the surface. In this form, the law was known as Gauss's Law.

In 1799, the Italian physicist chemist Alessandro Giuseppe Antonio Anastasio Volta (1745–1827) invented the electric battery. His battery consisted of a stack of copper and zinc plates separated by an electrolyte, which is a medium that is electrically conducting through the movement of ions but not conducting electrons. By measuring the voltage between two charged conducting plates, separated by an insulating layer, called a capacitor, Volta investigated electrical capacitance. He developed the means to study the relationship between electrical charge and voltage, showing they were proportional. His research led to Volta's Law of Capacitance. The results of his early work were sent to the Royal Society in 1800. Referred to as a Voltaic Pile, it substantially facilitated the study of electric phenomena. Volta was also credited for discovering and isolating methane, igniting it by an electrical spark.

In experiments using the Voltaic Pile in England by William Nicholson (1753–1815) and Anthony Carlisle (1768–1840) to decompose water into hydrogen and oxygen, electrolysis, a process in which an electric current causes a chemical reaction, was discovered. These studies led to the field of electrochemistry.

Magnetism

In contrast to the aforementioned laws of electrostatics, the basic laws of a magnetic field did not follow easily from earlier studies of magnetic materials. It was slow in developing because of a major difference between magnetostatics and electrostatics. There are no free magnetic charges as there are free electric charges in electrostatics. In magnetostatics, the basic element in magnetic materials is the magnetic dipole, which is determined by the molecular structure of the material.

In 1820, the Danish physicist chemist Hans Christian Ørsted (1777–1851) published the observation that electric currents create magnetic fields. His discovery was accidental. When he inadvertently switched off the current flowing in a coil of wire that was close to a nearby magnetic compass, he observed the needle deflected from magnetic north for an instant. This movement was the first connection found between magnetism and electricity. His investigation showed that electric current flowing through a wire generated a magnetic field that surrounded the wire.

Later that year, the French physicist mathematician André-Marie Ampère (1775–1836), motivated by Ørsted's discovery, continued with extensive experiments and analyses. Ampère demonstrated that two parallel wires with electric currents attract or repel each other, depending on whether the currents flow in the same or opposite directions, respectively. He observed that the mutual attraction or repulsion of two segments of current carrying wire was proportional to their lengths and the intensities of their currents. He found that a current-carrying wire produced a vector magnetic field B perpendicular to and surrounding the wire. From these observations, Ampère developed the fundamental law relating the strength of the magnetic field B to the amount of electric current I carried by the wire.

The magnetic field B was analogous to the electric field E in electrostatics. Specifically, Ampere showed that the basic vector connection relating B to I was the following: let dl be an element length of a thin wire that carries a current I and x the coordinate vector from dl to an observation point P, then the elemental magnetic field vector dB at P was given in direction and magnitude by $dB \sim k\,I\,(dl\ X\ x)/x^3$. The symbol X was the cross product of the two vectors $(dl\ and\ x)$, which generated the third vector dB perpendicular to the two original vectors dl and x.

The relation was given the name Ampere's Law and was essentially an Inverse-Square Law similar to Coulombs Law of electrostatics and Newton's Law of gravitation. The magnetic field produced by a current carrying wire behaved in a similar way to the force fields emanating from electric charges and gravitational masses. In 1827, Ampère published his research titled *Mémoire sur la théorie mathématique des phénomènes électrodynamiques uniquement déduite de l'experience*. Electrodynamics became the name of the new science.

In 1831, the English scientist Michael Faraday (1791–1867) made the first quantitative measurements relating time-dependent electric and magnetic fields on the action of currents placed in time-varying magnetic fields. He observed that a transient current was induced in a circuit if: (1) a steady current flowing in an adjacent circuit was turned off or on, (2) the adjacent circuit with a steady current flowing was moved relative to the first circuit and (3) a permanent magnet was pushed into or out of the circuit. No current flows unless there was either relative motion or the adjacent current changes.

Faraday interpreted the transient current flow as resulting from a changing magnetic field B linking the circuit. His observations could be summed up in the mathematical law $V \sim dB/dt$, where V was the voltage induced in the circuit and B was the magnetic field linking the circuit. He used the concept of lines of force or flux, to illustrate the magnetic field, surrounding a magnet or a conductor carrying a direct current and established the concept of the electromagnetic field. His discoveries and inventions of electromagnetic rotary devices expedited the electric motor.

As in thermodynamics, the scientists instrumental in clarifying the concepts of electricity and magnetism were honored by having their names attached to the electrical and magnetic quantities and their units. In the modern system of units, the electric potential is in volts (V), the unit of electric charge is the Coulomb (C), the electrical current flow is the ampere (A) defined as one Coulomb per second, the magnitude of magnetic field (in cgs [centimeters, grams, seconds] units) is the gauss (G) and the magnitude of capacitance (F) is a farad defined as one coulomb per volt.

Electromagnetic Waves

In 1865, the Scottish physicist James Clerk Maxwell (1831–1879), combined the preceding physical laws and observations of electricity and magnetism into a coherent mathematical framework of four equations. These equations correspond to descriptions of the following:

(1) Gauss's Law for electric fields, states that the flux of an electric field leaving a volume is proportional to the amount of electric charge within that volume,
(2) Gauss's Law for magnetic fields, states that there can be no net flux of magnetic field exiting of any volume,
(3) Ampere's Law, states that if an electric current passes through a surface, a magnetic field will circulate on that surface, with a magnitude proportional to the strength of the current, and
(4) Faraday's observations that electric currents flow in a magnetic field only if the magnet is accelerated.

Faraday's observations showed how electric fields were produced when the magnet moves. In contrast, Ampere's Law ignored the possibility of a time-dependent electric field to produce a magnetic field, as between the plates of a capacitor in an electric circuit. For example, when an electric current is stopped by a capacitor, positive and negative charges accumulate on the two separate plates of the capacitor and an electric field builds up between the plates. Maxwell's insight was to add a term reflecting this possibility to Ampere's Law. His insight was that the laws of electric and magnetic fields should have a mathematical similarity. As a result of Maxwell's mathematical model, the fact that light is an electromagnetic wave was discovered and explained.

These verbal descriptions (with the addition to Ampere's Law) are written as the following quartet of partial differential vector equations, known as Maxwell's Equations. They are expressed as:

$$\nabla \cdot \boldsymbol{D} = 4\pi\rho \qquad (10.1)$$

$$\nabla \cdot \boldsymbol{B} = 0 \qquad (10.2)$$

$$\nabla \times \boldsymbol{H} = 4\pi \boldsymbol{J}/c + 1/c\, \partial \boldsymbol{D}/\partial t \qquad (10.3)$$

$$\nabla \times \boldsymbol{E} = -1/c\, \partial \boldsymbol{B}/\partial t \qquad (10.4)$$

where $\boldsymbol{D} = \epsilon \boldsymbol{E}$ and $\boldsymbol{B} = \mu \boldsymbol{H}$, ϵ and μ are properties of the materials and the symbols ρ and $\boldsymbol{J}$ are the charge and current density, respectively. In this form the beauty of their symmetry is illustrated.

The last term in the third equation, $1/c\, \partial \boldsymbol{D}/\partial t$, was labeled by Maxwell the displacement current. It was analogous to $1/c\, \partial \boldsymbol{B}/\partial t$ in the last equation and was of central importance. Without it there would be no electromagnetic radiation. By combining the two curl equations (Eqs. 10.3 and 10.4) and making use of the vanishing divergences (in the absence of sources), one sees that each component of $\boldsymbol{E}$ and $\boldsymbol{B}$ propagate through space as a wave with the velocity c independent of frequency. Further manipulation demonstrates that $\boldsymbol{E}$ and $\boldsymbol{B}$ are both in phase, perpendicular to the direction of propagation and separated by 90^0 in their common plane. The wave created is labeled a transverse wave.

Maxwell noted that the speed of these electromagnetic waves was independent of frequency and was unusably close to the measured speed of light 3×10^8 m/s. He made the seminal step that light was an electromagnetic wave. He suggested that other electromagnetic waves, of all frequencies, could be produced traveling at the same speed as light. Infrared and ultraviolet rays were already known to exist. They were part of the same phenomena. Maxwell thus unified a previously unrelated range of observations. Maxwell's work was published in his definitive *Treatise on Electricity and Magnetism* in 1873. A decade later, the British mathematician Oliver Heaviside (1850–1925) reduced Maxwells original 20 equations to the elegant vector calculus equations illustrated above.

Maxwell's theory suggested electromagnetic waves could be produced by rapidly changing a current, as happens in certain electrical circuits excited by sparks. In 1887, the German physicist Heinrich Hertz (1857–1894) demonstrated that sparks create electromagnetic waves. He confirmed that these waves reflect from metal surfaces and refract in insulating materials just as does reflection and refraction of light and heat waves (infrared radiation). Hertz's research established conclusively that light and heat waves were electromagnetic radiation. Later, his papers were translated into English and published in three volumes: *Electric Waves* (1893), *Miscellaneous Papers* (1896) and *Principles of Mechanics* (1899). Named in his honor, the unit of frequency (cycles per second) is the hertz (Hz).

The expanding electronics industry, with its myriad of applications, can credit Maxwell's equations as its origins. Maxwell, along with Sir Isaac Newton, was ranked among the scientists who have had the greatest influence on twentieth century physics for the fundamental nature of their contributions. His quartet of equations to the field of electrodynamics is viewed as similar to Newton's laws of motion were to the field of mechanics. Albert Einstein is quoted as saying "I stand not on the shoulders of Newton, but on the shoulders of James Clerk Maxwell."

Additional Important Discoveries of the Nineteenth Century

Discoveries in the nineteenth century were prolific, not only in physics and chemistry, but also in the other fields of science, specifically, the life sciences. The majority had a profound effect on civilization.

In 1859, the English geologist, biologist and naturalist Charles Darwin (1809–1882) published his theory of evolution, with irrefutable evidence in *On the Origin of Species.* This evidence included material he collected on a five-year Beagle's survey voyage made in the 1830s. During the long voyage, he was greatly influence by the writings of Scottish geologist Sir Charles Lyell's (1797–1875), conception of gradual geological change, published in *Principles of Geology* (1830–1833), as his own observations also confirmed those of Lyell.

Darwin's publication introduced the concept that all species of life descended from common ancestors over time. His theory of natural selection demonstrated that the struggle for life resulted in life forms, which were similar to those man-made selections involved in breeding. Life less suited to the environment was less likely to survive and less likely to reproduce; life more qualified to the environment was more likely to survive and reproduce, giving their inheritable traits to future generations. This repetitive process over time produced life able to adapt to new environments. These variations accumulated to form new species. Darwin's concept of evolutionary adaptation through this natural selection process became primary to evolutionary theory. It is considered the unifying concept of the life sciences.

In 1861, the French microbiologist and chemist Louis Pasteur (1822–1895) published his germ theory proving that germs caused disease. He developed vaccines for rabies and anthrax, and invented the process of pasteurization. Pasteurizing milk improved diets and food became safer. In 1887, he founded the Pasteur Institute.

The idea of the germ theory was promoted by the German physician and microbiologist Robert Koch (1843–1910). He provided experimental support for the concept of infectious disease, using animals and humans. He identified the causative agents of anthrax, cholera and tuberculosis. In 1905, Koch received the Nobel Prize in Medicine for his research on tuberculosis.

The work of Pasteur and Koch led to a substantial decrease in death from infection and accounted for a significant increase in life expectancy in Europe, during the later part of the nineteenth century.

Chapter 11
Physics: Around 1900

William Thomson's contributions to physics, including the second and third laws of physics, were countless, advancing the understanding of thermodynamics and electromagnetism. The Kelvin temperature scale named after him established the temperature of -273 °C at a value of zero degrees Kelvin. Throughout his life he had an over riding goal of making science useful. His effort was apply illustrated with his involvement in telegraphy and efforts to advance the transatlantic cable projects in 1857–1858 and 1865–1866. This work earned him the aristocratic title of Lord Kelvin and entry into the House of Lords in 1892.

So when Lord Kelvin made his famous statement in his address to the British Association for the Advancement of Science in 1900: *There is nothing new to be discovered in physics now. All that remains is more and more precise measurement.*, it had authority and was widely accepted as the truth. Indeed, the majority of the phenomena known up to that time fit into Newton's theory of mechanics or the electromagnetic theory of Maxwell almost exactly. This world view, however, was about to be changed, during the end of the nineteenth century and the beginning of the next. Up to then, matter was considered to be composed of unchangeable structureless atoms and light was believed to be a continuous wave of oscillating electric and magnetic fields. During the nineteenth century, technology improved significantly and curiosity motivated enhancing the limits of measurement to investigate these assumptions.

X-Rays

In 1895, the German physicist Wilhelm Roentgen (1845–1923) discovered x-rays while investigating the phenomena of gaseous discharge in cathode-ray tubes. These tubes were sealed cylinders of glass in which the bulk of the air was evacuated. There, Roentgen applied a high voltage between two electrodes interior to the tube,

T. Sanford, *A Whirlwind History of the Universe and Mankind*,
https://doi.org/10.1007/978-981-97-2674-5_11

located at either end. Within the tube, tens of kilovolts was applied between the negative electrode (cathode) and the positive electrode (anode). The voltage caused a beam of cathode rays (particles) to accelerate from the cathode to the anode, which escaped through an aperture in the anode. The escaping particles formed a beam, which impacted the far end of the tube. The particles were detected using phosphors painted on the tube end, where they emitted light when impacted by the beam.

With the tube covered in black cardboard, Roentgen discovered a fluorescent glow in a photo-sensitive screen three meters away. Investigating, he found that the fluorescence was caused by the accelerated cathode particles hitting the glass at the end of the tube. Further exploring, demonstrated that a new kind of ray produced by the cathode particles hitting the tube glass could pass through many materials, including soft tissues of the body but not bones or metals. He observed that the rays could be revealed by the darkening of wrapped photographic plates as well as by fluorescence in various materials. Because of the unknown nature of the rays, with their great penetrating power, he named them x-rays.

In January, 1896, Roentgen gave his first public presentation *On a New Kind of Rays* at the Würzburg Physico-Medical Society. In closing the lecture with a demonstration, he made a photographic plate of the hand of an attending anatomist. Few physics discoveries have had such an immediate impact as Roentgen's x-rays image did. Within a year of his announcement and demonstration, the application of x-rays to diagnose, evolved to become a significant part of the medical profession. In 1901, Roentgen was awarded the first Nobel Prize in physics "...in recognition of the extraordinary services he has rendered by the discovery of the remarkable rays subsequently named after him."

Radioactivity

In 1883, the French physicist Henri Becquerel (1852–1908) began studying fluorescence (the emission of light by a substance that has absorbed light or other electromagnetic radiation), and phosphorescence (the emission of radiation in a similar manner to fluorescence but on a longer timescale, so that emission continues after light excitation ceases), following in the research of his father and grandfather. He was particularly interested in the phosphorescent of uranium salts. Hearing about Roentgen's x-rays in January 1996, he thought perhaps there might be a connection between x-rays and radiation from uranium salts in which phosphorescent uranium salts might absorb sunlight and reemit it as x-rays.

To test this idea, Becquerel designed an experiment in which metal cutouts were placed between crystals of uranium and photographic plates wrapped in black paper to keep out light. It was then placed in sunlight to expose the uranium crystals. When developed, the photographic plates showed outlines of the metal cutouts and the crystals, supporting his conjecture. Thinking his hypothesis that sunlight caused phosphorescence was confirmed, he reported his results to the French Academy of Science meeting the following month. In a second experiment, using the same setup that had been placed in a darkened drawer and not exposed to sunlight, he developed the stored away plates anyway; only to find similar images of the cutouts and crystalline structure. In subsequent experiments, Becquerel observed that sun light was not needed and that uranium compounds that were non-phosphorescent also showed the same effect. Realizing the importance of his discovery, in May 1896, Becquerel announced, that it was the element uranium itself that was spontaneously emitting the radiation.

Two years later, in Paris, Marie (1867–1934) and Pierre Curie (1859–1906) began investigating the new uranium rays. They found ways to measure the intensity of the radioactivity, named by Marie Curie to describe the new phenomena. Other natural radioactive elements like thorium, polonium and radium were soon discovered.

For Becquerel's discovery of spontaneous radioactivity, he was awarded half of the 1903 Nobel Prize for Physics. The other half was shared between Marie and Pierre Curie for their study of the Becquerel radiation. Marie Curie became the first woman to win the Nobel Prize in Physics. In 1910, she succeeded in isolating pure metallic radium, for which she received a second Nobel Prize, in Chemistry in 1911.

Although late in studying radioactivity, Ernest Rutherford (1871–1937) from New Zealand, became dominant in the field and would be referred to as the father of nuclear science. In 1884, Rutherford received a scholarship to work under the English experimental physicist J. J. Thomson (1856–1940) and moved to Cambridge University, England, for post-graduate studies at Cavendish Laboratory. Rutherford's initial research led to investigations of electricity and radiation and later to a detailed study of radioactivity. By 1898, Rutherford discovered the existence of alpha, beta and gamma rays in uranium radiation, identifying many of their characteristics. That year he left Cavendish to become Macdonald Professor of Physics at McGill University in Montreal, Canada.

While at McGill, Rutherford continued his research into radioactivity, where he discovered that some heavy elements decay into lighter elements over fixed periods of time. Working with others, he determined that radioactivity is due to the decay of the radioactive element, transmuting into a daughter element; and that each type of element has its own transformation period or half-life. Radioactive dating depends largely on this fact. This discovery motivated him and his colleagues to search for other radioactive elements and document their decay series. In 1904, Rutherford published his first book on the topic, *Radioactivity*. Four years later in 1908, he was awarded the Nobel Prize in Chemistry for his research on the transmutation of elements and the chemistry of radioactive substances.

The Electron

In 1897, J. J. Thomson discovered the electron indirectly by measuring the charge-to-mass ratio of the cathode particles generated in low voltage Cathode Ray Tubes. In contrast, and in addition to Roentgen's configuration, Thomson placed two oppositely charged electric plates, exterior and around the tube, downstream of the anode and tube end to examine characteristics of the particles in a low voltage electric field.

The particles within the tube were deflected away from the negatively charged plate and towards the positive plate, indicating that the particles were of negative charge. By placing two external magnets, in similar fashion, of opposite polarity on either side of the tube the characteristics of the particles were further explored in the resulting magnetic field. Using the classical laws of mechanics and electromagnetism, Thomson determined the charge-to-mass ratio of the cathode particles.

These measurements led to a fundamental discovery that the mass of each particle was significantly smaller (orders of magnitude smaller) than that of any known atoms. The experiments were repeated using cathodes of differing metals. The results remained constant independent of what cathode material they used. As a result of his experiments, Thomson concluded: (1) the cathode particle is negatively charged, (2) the particles must exist as part of the atom as the mass of each particle was only 1/2000 of the mass of a hydrogen atom and (3) these particles were found to be in all atoms. In 1897, it was suggested that these new particles be given the name electrons.

Initially controversial, Thomson's discovery disproved Dalton's atomic theory, which assumed that atoms were incapable of division. An entirely new model for the atom was needed to account for the existence of electrons. Knowing that atoms were neutral led Thomson to propose that atoms could be described as electrons floating in a sea of diffuse positive charge. This model became known as the plum pudding model of the atom.

Thomson's later work with positively charged particles led to the development of the mass spectrograph. He received the Nobel Prize in Physics in 1906 "...in recognition of the great merits of his theoretical and experimental investigations on the conduction of electricity by gases." In 1909, he was knighted.

In 1909, the American experimental physicist Robert Millikan (1868–1953), began experiments to measure the electrical charge of the electron with greater accuracy than Thomson. His idea was to use oil drops in his now famous oil drop experiment. The experiments measured the behavior of the drops with variable electoral charges falling (in the Earth's gravitational field) or being accelerated between two voltage carrying plates. The kinetic energy of the drops was established by measuring the potential energy of the electric field needed to stop them. His measurements proved that the charge on the drops were always whole numbers in multiples of 1.60×10^{-19} C. The charge was no longer a continuous quantity but discrete. The minimum charge provided the basic unit of charge for the electron. His results were published in 1913.

Quantum Structure

Blackbody Radiation

Understanding the physics of radiation was problematic late in the nineteenth century. The surface of every physical body at a temperature above absolute zero emitted electromagnetic radiation. This process was viewed as the result of the acceleration of electric charges near the surface due to thermal agitation of atomic oscillators. From the many different acceleration processes that gave rise to thermal radiation, a wide spectrum of wavelengths was assumed to be produced.

Thermal emission from the interior of a cavity through a small hole in the cavity was referred to as blackbody radiation and had the same spectrum as radiation emitted from any black surface. The radiated energy E from the hole, over all frequencies, was found to be proportional to the temperature of the cavity T raised to the fourth power, independent of cavity material: $E \sim T^4$. This relation became known as the Stefan-Boltzmann Law. Austrian physicist Josef Stefan (1835–1893) formulated the Law in 1879; in 1884 it was derived by Boltzmann based on thermodynamic principles.

The material independence of the Stefan-Boltzmann Law suggested that other properties of the blackbody radiation, such as its spectral distribution, may depend only on the temperature and not on the material. This proved correct and the universal nature of the blackbody emission thus drew significant theoretical interest.

A number of well-known physicists advanced theories based on classical physics to explain the spectra from a blackbody source. All had limited success. Famous among them were two English physicists, Lord Rayleigh (1842–1919) and Sir James Jean (1877–1946). They formulated what became known as the Rayleigh-Jeans model in 1900. In the limit of long wavelengths, their spectra approached experimental results, but became unrealistic at short wavelengths. This discrepancy became known as the "...ultraviolet catastrophe."

During this time, the German theoretical physicist Max Planck (1858–1947) heuristically derived an explanation of blackbody radiation that fit the measured spectra precisely. He assumed two radical assumptions with regard to the atomic oscillators:

(1) The oscillators cannot have any energy E but instead were restricted to those given by $E = nhf$ where n was a integer (now called a quantum number), h was a constant (now called Planck's constant) and f was the oscillator frequency, and

(2) The oscillators do not radiate energy continuously but only in discrete jumps (now called quanta).

These quanta were emitted only when the oscillators made transitions between energy states. If an oscillator remained in one of those states, it neither emitted nor absorbed energy. If n changed by one, then the amount of energy radiated was $\Delta E = hf$. Once emitted from the cavity walls, the radiation was still in the form of an electromagnetic wave. Because the value of Planck's constant h was very small, namely, 6.63×10^{-34} J-s, the discreetness in the energy of the waves was not detectable.

Planck's formula is given by $R = 2\ h\ c^2\ /_\lambda{}^5\ 1/(e^{\ (h\ c/\lambda\ kT)} - 1)$, where R was the spectral radiancy defined as the rate at which energy is radiated per unit area as a function of wavelength λ, c is the speed of light, and h is Planck's constant, T is the wall temperature and k the Boltzmann's constant. Examination of the equation indicates that, for small wavelengths, the radiancy decreases with decreasing λ, thus avoiding the "...ultraviolet catastrophe."

In December 1900, Planck presented his theory of blackbody radiation to the Berlin Physical Society. Quantum physics theory dates from that moment and revolutionized the future understanding of atomic and nuclear matter. Planck was awarded the Nobel Prize for laying the foundations of quantum mechanics in 1918. In 1948, the Kaiser Wilhelm Society (founded in 1911) was renamed the Max Planck Society for the Advancement of Science in recognition of Planck's scientific accomplishments. In 2023, this Society includes 83 different scientific institutions in Germany.

It was the independent observation of the predicted blackbody CMB (Cosmic Microwave Background) in 1960 that gave the Big Bang its credibility. In 1989, the COBE (Cosmic Background Explorer) satellite measured the CMB spectrum from 0.5 to 10 mm, over its full spectral range, which peaked at approximately 1 mm. The measurements demonstrated definitively that the above Planck equation for blackbody radiation precisely fit the COBE spectrum with a temperature of 2.73 degrees Kelvin. These measurements provided both independent credibility to the Big Bang and subsequently to Planck's analysis itself.

Photoelectric Effect

In 1887, Heinrich Hertz, during his research on generating electromagnetic waves, discovered that if metal electrodes were illuminated with ultraviolet light, the voltage needed to make a spark jump between electrodes in his spark-gap generator was reduced. In 1899, J. J. Thomson demonstrated that ultraviolet light striking a metal surface caused the ejection of electrons. These physicists were among the first to observe the photoelectric effect.

In 1902, quantitative measurements of the photoelectric effect were made in experiments conducted by Hungarian-born German Philipp Lenard (1862–1947). Together with others he established that: (1) metal plates irradiated with light may emit electrons, named photoelectrons, creating a photoelectric current, but also that (2) for a given photosensitive cathode, there was a critical frequency f_0 of the light, below which nothing happens, independent of how strong the light intensity is, (3) the magnitude of the photo current was directly proportional to the intensity of the light for $f > f_0$, (4) the energy of the photoelectrons increased linearly with the frequency f of the light and (5) the energy of the photoelectrons was independent of the intensity of the light.

In 1905, Lenard received the Nobel Prize "...for his work on cathode rays and the discovery of many of their properties," which was published in *Uber Kathodenstrahlen* the following year. The properties of the photoelectrons, however, remained an enigma especially characteristics (4) and (5). These were not easy to understand, considering that light was thought to be a wave.

In the same year, the German-Swiss theoretical physicist Albert Einstein (1879–1955) postulated a bold explanation: light comes in packets of energy. Energy in the light beam travels through space in concentrated packets, called photons. The energy E of a single photon was given by $E = hf$, where h was Planck's constant and f was its frequency.

Previously, Planck assumed that light traveled through space as an electromagnetic wave, although in blackbody radiation it was emitted discontinuously from a source. In contrast, Einstein suggested that light traveling through space behaves not like a wave, but like a particle. Einstein proposed that the kinetic energy of the electron escaping from the photocathode would be equal to $E = hf - W$, where W was the work function of the cathode and depends on the material of the cathode. It was the energy required to overcome the attractive forces present at the surface, which kept the electrons from escaping. Einstein's explanation was published with the title *On the Electrodynamics of Moving Bodies* in *Annalen der Physik* in 1905.

In 1916, Millikan carefully measured the maximum kinetic energy E of the photoelectrons as a function of the generating light frequency f in discrete steps between 6 and 12×10^{14} s^{-1}. The precision data exhibited a perfect linear dependence with frequency, in accordance with the above Einstein equation, with the slope $\Delta E/\Delta f$, giving exactly the value of Planck's constant h of 6.57×10^{-34} J-s to a precision of 0.5%.

Millikan had made the first direct photoelectric determination of Planck's constant h superior to any made previously. These measurements put the quantization of light energy on a firm footing. In 1922, the Nobel Prize was awarded to Einstein for "...his services to theoretical physics, and especially for his discovery of the law of the photoelectric effect." Millikan was awarded the Nobel Prize in 1923 "...for his research on the elementary charge of electricity and the measurements of the photoelectric effect."

Space and Time

Special Theory of Relativity

In 1887, the speed of light c was measured by the Americans Albert Michelson (1852–1931) and Edward Morley (1838–1923) to be the same, independent of the motion of the source. The measurements included those of the Earth relative to the Sun in its orbit around the Sun at 30 km/s or 1/10,000 of the speed of light. These measurements were accurate enough to notice the 0.01% change that would occur if the orbital speed were to add or to subtract from the light speed value measured at rest: 300,000 km/s.

In Newtonian mechanics, the constancy of the measured speed was different from that expected by experience. The anticipated speed of any moving object relative to its own speed was obtained by adding or subtracting the measured speed of the object's speed to its own speed. This contrast between measurement and expectation needed understanding.

In 1905, Einstein published a paper titled *On the Electrodynamics of Moving Bodies* in the German Journal *Annalen der Physik* that provided an explanation of Michelson's and Morley's data that went beyond classical Newtonian physics. Thanks to Planck's acceptance of Einstein's ideas, the paper generated a profound effect on the conception of space, time, mass and energy.

Einstein's analyses, based on *gedanken* (thought) experiments, referred to how one observer would describe that which another would experience when moving at a constant speed relative to the first, using light beams. The situation could be between an observer stationary on the ground beside the other on a moving train. For these inertial frames, defined as free of acceleration and gravitational fields, Einstein derived the Laws of the STR (Special Theory of Relativity).

Einstein based his laws on two postulates: (1) the laws of mechanics were the same in all inertial frames, despite the fact that these frames may be in uniform motion with respect to each other and (2) the velocity of light was independent of the motion of its source. The second postulate was motivated by both Maxwell's equations and the Michelson-Morley measurements of the speed of light.

Einstein's theories regarding time concluded that if two observers were measuring the same speed of light c even though one was moving relative to the source of light, what was changing was not the speed per second but the second itself. Time, assumed to be a constant everywhere in space, was not constant at all. For an observer who was moving, the second was longer and thus the speed remained the same for either observer. Time became a forth dimension in a three-dimensional universe forming a four-dimensional space–time.

Specifically, the time difference Δt observed in a stationary frame between two events occurring at the same place in a moving frame of velocity v, derived in the STR, was given by $\Delta t = \Delta t'/(1 - v^2/c^2)^{1/2}$. The time difference $\Delta t'$ was the quantity measured by the observer in the moving frame. When v was small, with respect to the speed of light c, the difference between which frame the event was measured in becomes negligible: $\Delta t \sim \Delta t'$. When the velocity approaches the speed of light c in the moving frame, however, the denominator becomes small and Δt increases. The increase in Δt measured by the observer in the stationary frame was referred to as time dilation. Moving clocks were predicted to tick more slowly than stationary clocks. The quantity $1/(1 - v^2/c^2)^{1/2}$ is referred to as the Lorentz factor (previously identified by the Dutch physicist H. Lorentz), which only becomes large when velocities approach that of light.

The STR relations, illustrated that time and space were not separate, but that space–time must be viewed as having four mutually dependent dimensions of reality. Time t connects with space x when there is movement between inertial frames, as in the measurement of velocity v when motion is involved: $v = dx/dt$. The STR relations dissolve into the old Galilean transformations (relations) known since the time of Galileo when the velocity between frames becomes small (with respect to c). Velocity then can be added or subtracted to predict a speed.

To account for the experimental fact that the speed of light c was measured to be the same in all inertial frames, the STR predicted other new physics results in addition to time dilation, *i.e.*, length contracted and mass increased as the speed of a body increased. Additionally, no object or field could travel faster than the speed of light and significantly, the equivalence of mass m and energy E, leads to the famous Einstein equation $E = mc^2$.

The STR did not achieve full acceptance, until its predictions had been proven. One of the first of these tests was made in 1909 by the German physicist Alfred Bucherer (1863–1927). Bucherer measured the mass of high velocity electrons in experiments similar in concept to those designed to measure the charge-to-mass ratio of low velocity electrons by J. J. Thomson. A radium source was placed at the center of a circular capacitor consisting of two silvered glass plates spaced 0.25 mm apart that were charged to 500 V and set in a uniform 150 Gauss magnet field. The radium emitted high velocity electrons in all directions. Only those that exited the gap between the plates were those whose speed the electric and magnetic field exactly compensated one and other. After leaving the capacitor, the electrons were deflected by the magnetic field and exposed a photographic plate set perpendicular to the capacitor's rim. Over a range of many different electron velocities v, with v/c measurements ranging from 0.3 to 0.8, the measured mass of the electron dilated from 1.05 to 1.65. This was in exact agreement with the STR equation of mass versus velocity.

One of the purest verifications of this theory involved the measured time dilation of the lifetimes of very fast cosmic muons (which have a half-life of 2.2 μs when at rest) produced in the upper atmosphere. In 1963, the American physicist James Frisch (1918–1991) compared the number of comic ray muons of velocities generated between 0.995 c and 0.9954 c measured atop Mount Washington in New Hampshire to those measured at ground level at Cambridge, Massachusetts. The vertical distance and flight times between the two locations were 1,907 m and 6.4 μs, respectively. At the mountain top, 563 muons per hour were measured. If there was no time dilation in their life-times, only 27 muons per hour would be measured at the ground. Instead, 412 muons per hour was measured. The number resulted in a measured time dilation of 8.8 ± 0.8, in agreement with the STR prediction.

It was not until the discovery of nuclear physics, near the half-century mark, that the significance of the STR relations became important, when mass was converted to energy. Moreover, the use of the STR equations became essential for particle physicists when predicting the motion of particles that travel close to the speed of light in modern day accelerators.

In 1907, Michelson became the first American to win the Nobel Prize in Physics "…for his optical precision instruments and the spectroscopic and metrological investigations carried out with their aid." illustrating the important impact of his earlier measurements.

General Theory of Relativity

In 1915, Einstein presented his GTR (General Theory of Relativity) to the Prussian Academy of Science. GTR went beyond the STR to consider motion in accelerating frames of reference and defined the unique role that gravity plays in the universe. STR became a special case of GTR.

Einstein was motivated, in part, by the equivalence of inertial and gravitational mass. Since, inertial mass m_i was the mass that appeared in Newtons second law $F = m_i\,a$ and gravitational mass m_g (and Mg) were the masses that appeared in Newton's universal law of gravitation $F = g\,M_g\,m_g/r^2$, that accounted for the force of attraction between massive bodies, there was no a priori reason why the two masses (inertial and gravitational) should be identical. GTR embodied this equivalence in its development, leading to Einstein's conclusion that there was no difference between the force on a body in a gravitational field and that experienced by a body undergoing acceleration. The equivalence principle states being at rest in a gravitational field or being in accelerated motion were physically identical.

To resolve this equivalence and other issues, Einstein proposed that space–time was curved. He used the non-Euclidean mathematics of Riemannian Geometry to develop field equations that connected the curvature of space–time to matter, energy and momentum. Essentially, GTR was a theory of gravitation whose defining feature was its use of non-linear field equations, where the solutions were metric tensors that defined the four-dimensional topology of the space–time and how objects move within it.

The solutions made a few minor corrections to the results of Newton for the solar system and correctly explained the anomalous perihelion advance of Mercury about the Sun. Mercury, being the closest planet to the Sun, felt the curvature more than those at a greater distance. GTR had its most significant impact, however, when it predicted the bending of light in strong gravitational fields. In 1919, the English polymath Sir Arthur Eddington (1882–1944) observed the bending of star light by the gravitational field of the Sun, during a total eclipse off the west coast of Africa, which resulted in making Einstein famous.

In addition to the bending of light when passing a massive object, one of the other predictions was that of time dilation (clocks tick slower) in gravitational wells. With the advent of precision cesium clocks, the prediction of gravitational time dilation, was confirmed in a variety of experiments. In 1976, for example, L. Briatore and S. Leschiutta compared the rates of two identical cesium clocks, one at 250 m above sea level in Turin, Italy and the other at the Plateau Rosa, Switzerland 3,500 m above sea level. Using two independent techniques, they found differences of 33.8 ± 6.8 ns/day and 36.5 ± 5.8 ns/day. Within the uncertainties, both differences were in exact agreement with the expected 30.6 ns/day from the GTR. The results were published in *II Nuovo Cemento B.* **37** (2): 219–231, 1977.

The relevance of the STR and GTR in today's world is best illustrated by the effect that speed and gravity have on clocks used in the GPS (Global Positioning System). These satellites operate at elevations of 20,200 km above the Earth's surface with speeds of 3.9 km/s. In orbit, a satellite clock loses 8.4×10^{-11} s for each second that passes due to its orbital speed (time dilation using STR). In contrast, 5.3×10^{-10} s for each second that passes was gained due to its altitude above the Earth (gravity effect from GTR). Putting both effects together, the time gained during the length of a 24-h day by a clock in a satellite, relative to a clock on the ground was approximately 40 μs. The time difference may not seem significant but when multiplied by the speed of light (which is done when evaluating the location of an object on the Earth) it corresponds to a distance of 12 km. Therefore, the GPS would become useless for providing any location to an accuracy better than 10 km, unless relativistic and gravitational effects are both included.

It was not until the early twenty-first century, however, that two of the most remarkable predictions of the GTR were to be made, namely, the existence of black holes and the existence of gravitational waves.

Chapter 12
Physics: 1900–1950

Atomic Structure

Rutherford Model

In Canada, Rutherford feeling isolated from the scientific centers of Europe, in 1907 accepted the Chair of Physics at Manchester University in England. With the German physicist Hans Geiger (1882–1945), they conclusively determined that alpha rays were doubly ionized He atoms. (Geiger eventually became known as the inventor of the Geiger Counter, the instrument they used to measure the alpha radiation.) Rutherford correctly assumed that because the alpha particles were more massive and of atomic dimensions compared to electrons, they would be key to understanding the nature of matter. His intuition was right.

In 1909, Rutherford made his greatest discovery, using alpha particles to strike thin (10^{-4} cm thick) gold foils. This experiment employed a radioactive uranium source emitting alpha particles encased in a lead shield. The alpha particles were collimated into a narrow beam by passing them through a slit in a lead shield before striking a target foil. Surrounding and behind the foil was a movable screen coated with zinc sulfide to render it fluorescent when impacted by the scattered alpha particles. The resulting scintillation was visible through a microscope that viewed the rear of the screen.

Based on an earlier plum-pudding-model of the atom that assumed the atom's charge and mass were uniformly distributed throughout the atom, Rutherford and others anticipated that the bulk of the alpha particles would pass through the foil with only a slight deflection, if any. This proved correct for most of the alpha particles, except for a small fraction that were deflected at large angles. Some were even directed backwards. In a famous comment, Rutherford exclaimed that it was "as if you fired a 15-inch shell at a piece of tissue paper and it came back and hit you." The number of alpha particles scattered at angles greater than a few degrees was significantly larger than predicted. Thus, a new model was required.

© The Author(s), under exclusive license to Springer Nature Singapore Pte Ltd. 2024   173
T. Sanford, *A Whirlwind History of the Universe and Mankind*,
https://doi.org/10.1007/978-981-97-2674-5_12

In 1911, Rutherford proposed that all the positive charge of an atom and all of its mass was concentrated in a small volume that he named the nucleus. Assuming this to be the case, Rutherford derived the angular distribution of scattered alpha particles expected, based on Coulomb's law and Newton's laws of Mechanics. The form of his angular distribution, illustrated by the proportionality, is

$N(\theta)\, d\theta \sim t\, (z\, Z/(m\, v^2/2))^2\, (sin(\theta)\, d\theta\, /sin^4(\theta/2))$. $N(\theta)\, d\theta$ is the number of particles scattered into the angle θ within the interval $d\theta$. The letter m represents the mass and v the velocity of the scattered alpha particles of charge z from a target of thickness t, having atoms of mass M and charge Z.

This scattering distribution was tested, using both gold and silver targets in the months following its derivation showing: (1) the measurements over the angular range 5–150° agreed with the above theoretical distribution to within a few percent, (2) $N(\theta)\, d\theta$ was proportional to the thickness t over a factor of ten, (3) the dependence of the inverse square of the kinetic energy dependence ($m\, v^2/2$) was observed over an energy variation of a factor of three, (4) within the accuracy of the experiment the scaling with Z was observed and (5) the electrons had negligible impact on the alpha's scattering.

By evaluating the distance of the closest approach, Rutherford was able to place limits on the size of the nucleus, which proved to be in the range of 10^{-12} cm. At the time, the size of atoms was known to be of the order of 10^{-8} cm. The electrons thus occupied the bulk of the empty space around the nucleus, with the nucleus being an 10,000 times smaller. An analogy would be to expand the atom to be the size of a football stadium, then the nucleus would be the size of a marble. This model of the atom became known as the Rutherford or Nuclear model.

Rutherford also noted that the measured charge of the gold nucleus was large, corresponding to approximately half the atomic weight of the atom in whole number units of hydrogen mass (today called atomic number). A month after reading Rutherford's 1911 paper, the Dutch physicist Antonius van den Broek (1870–1926) was the first to realize, that the atomic number of an element in the periodic table corresponds to the charge of its atomic nucleus. His hypothesis was published in 1911.

The English physicist Henry Moseley (1887–1915), inspired by van den Broek, proved the hypothesis was correct. By 1913, Moseley had found a systematic relation between the atomic number and the frequency of the spectra measured in many elements, using the technique of x-ray crystal spectroscopy. This relation is currently known as Moseley's Law. Had his life not been shorten by his participation in World War I (1914–1918) he would have likely received a Nobel Prize for his discoveries.

In 1914, Rutherford was knighted. In 1918, during a series of experiments that involved scattering alpha particles in nitrogen gas, Rutherford made another significant discovery. In the arrangement, some of the nitrogen changed into oxygen atoms and, hydrogen nuclei were generated. Based on Rutherford's earlier experiments in which hydrogen nuclei were generated when alpha particles were directed at hydrogen gas, Rutherford surmised that the hydrogen nucleus was part of the nitrogen before the impact. As a result, he theorized that hydrogen nuclei were the building material of nuclei, which he called protons. This process generated the first artificial splitting of an atom.

In 1919, Rutherford succeeded J. J. Thomson as head of the Cavendish Laboratory, Cambridge University and in 1931 he was elevated to a baron. In 1937, Lord Rutherford was laid to rest near Sir Isaac Newton and Lord Kelvin and a memorial stone for James Clerk Maxwell in Westminster Abbey. In 1957, the Rutherford High Energy Laboratory was opened, which was set next to the Atomic Energy Research Establishment in England. In 1992, Rutherford's picture was placed on the New Zealand $100 note. Among his many honors, was that the element Rutherfordium (Rf, atomic number 104) was named after him.

Rutherford's gold foil experiment became a classic template for future high-energy physics experiments. For example, almost sixty years later, the author of this book following in Rutherford's footsteps, scattered muons off of carbon nuclei to extract a precision measurement of the size of the proton, which had come into question. In the author's experiment at Columbia University's Nevis cyclotron, the cyclotron generated muons, analogous to Rutherford's alpha particles generated by decaying uranium, with the muons similar to the alpha particles scattering off a gold foil. Following publication of these measurements, this author also took a position at the Rutherford High Energy Laboratory.

Bohr Model

The Rutherford model of the atom was an important first step. However, it did not adequately address the nature of the negative electrons and the mechanism of how they filled the space surrounding the positive nucleus without falling into the nucleus. If the electrons filled the expanse similar to planets, rotating in orbits around the sun, being held in check by the electromagnetic force instead of the gravitational force, the electrons would radiate energy and spiral into the nucleus. According to electromagnetic theory, any charged particle moving on a curved path emits electromagnetic radiation and slows down.

In 1913, the Danish physicist Neils Bohr (1885–1962) solved this problem by proposing a number of postulates, similar to those proposed by Planck for blackbody radiation. His postulates were the following: (1) an electron in an atom moves in a circular orbit about the nucleus under the influence of the Coulomb attraction between the electron and the nucleus and obeys the laws of classical mechanics, (2) it was only possible for an electron to move in an orbit where its angular momentum L was an integral multiple n of Planck's constant h divided by 2π ($L = n\,h/2\pi$), (3) even though the electron was moving in an orbit it does not radiate energy and (4) radiation was emitted if an electron, initially moving in an orbit of total energy E_i, discontinuously changes its motion so that it moves in an orbit of total energy E_f. The frequency of the emitted radiation f was equal to the quanta ($E_i - E_f$) divided by the Planck's constant h [that is $f = (E_i - E_f)/h$].

Using the laws of classical mechanics together with the aforementioned assumptions, Bohr demonstrated that the frequency spectrum of an atom of charge Z could be expressed by $f = R\,Z^2\,(1/n_f^2 - 1/n_i^2)$, where n_f and n_i were integers and the constant R depends on the measured mass of the revolving electrons, their charge, the speed of light c and Planck's constant h. The early measured spectra of the one electron hydrogen atom all agreed exactly with the appropriate choice of n_f and n_i. These initially were the Balmer and Paschen spectra, named after the scientists who discovered them. Later, with differing values of n_f and n_i, the Lyman, Brackett and Pfund spectra were predicted and observed. The existence of spectral lines that had not been previously observed, before they were predicted by the Bohr model, gave justification to using the Bohr's model for the hydrogen atom.

This model also was successful when applied to the one-electron atom of singularly ionized helium. The predicted spectra was identical to that of hydrogen when setting $Z^2 = 4$ in the above expression for frequency. In general, the Bohr spectra for these simple atoms, when corrected for finite nuclear mass, agreed with the spectroscopic data to within a few parts in 100,000. In 1922, Bohr received the Nobel Prize in Physics for "…the investigation of the structure of atoms and of the radiation emanating from them."

The Bohr model assumed that the total energy of an atomic electron was quantized. That assumption was confirmed in the first electrical measurement by two German physicists James Franck (1882–1964) and Gustav Ludwig Hertz (1887–1975) in 1914. They designed a vacuum tube for studying energetic electrons that were accelerated through an anode grid of positive charge that were collected on a downstream plate of negative charge. The tube was filled with low pressure mercury gas. They discovered that when an electron collided with a mercury atom, it would lose only a fixed amount of its kinetic energy (4.9 eV) before disappearing. [1 electron volt (eV) $= 1.60 \times 10^{-19}$ J.] This measurement was revolutionary in that it disagreed with the belief that an electron could be held to an atom's nucleus with any amount of energy.

In a second 1914 paper, Franck and Hertz reported measurements of the light emission by the mercury atoms, which had absorbed energy from collisions with the electrons. They demonstrated that the frequency of the light corresponded exactly to the 4.9 eV of energy that the accelerating electron had lost. The relation between frequency and energy was central to the Bohr model. These measurements reinforced Bohr's assumptions. In 1925, Franck and Hertz were awarded the Nobel Prize in Physics "…for their discovery of the laws governing the impact of an electron upon an atom."

De Broglie Model

Planck's assumptions for the blackbody radiation and the atomic radiation from the Bohr Model had little justification except that when accepted they explained the measurements. In 1905 Einstein provided meaning for Planck's assumptions by proposing that light travels through space in concentrated energy packets (like particles), called photons. Each photon had a discreet frequency f given by $f = E/h$, with energy E quantized. In terms of the wavelength λ, the relation was expressed by

$\lambda = h\,c/E$. (Electromagnetic radiation travels at the speed of light c and its wavelength is $\lambda = c/f$.) Light thus had a wave particle duality.

In parallel to Einstein and the above relations, the French physicist Louis de Broglie (1892–1987), in his 1923 doctoral thesis, offered an explanation for Bohr's assumption of the restricted motion of electrons in atoms. He proposed that electrons (and particles, in general) propagate through space also like a wave, with a wavelength given by $\lambda = h/p$, where p is the electron's (or the particle's) momentum. In 1924, de Broglie reinterpreted Bohr's condition that the angular momentum L of the electron is an integer multiple of $h/2\pi$, as a standing wave condition. The atomic electron is a wave with a whole number of its wavelengths λ that fit along the circumference $2\pi\,r$ of the electron's orbit in the atom: $n\lambda = 2\pi\,r$. Inserting de Broglie's assumption $\lambda = h/p$, into this requirement, the previous equation becomes $nh/2\pi = r\,p = L$, which is Bohr's second postulate. (Note: a rotating electron's angular momentum about the nucleus at radius r is defined by $L = r\,p$.)

The electron wave was thus confined within boundaries imposed by the nuclear charge and was restricted in shape and motion. Any wave shape that does not fit within the atomic boundaries would interfere with itself and cancel out. In 1929, de Broglie received the Nobel Prize in Physics "…for his discovery of the wave nature of electrons."

At the time, there was no experimental evidence for de Broglie's concept that the particle characteristics of electrons, which were already established, might also show characteristics of the wave behavior of radiation. This would be analogous to light showing particle-like behavior. His intuitive thoughts were based on the symmetry of nature.

By 1927, two American physicists Clinton Davisson (1881–1958) and Lester Germer (1896–1971), provided the first confirmation of the wave nature of electrons in an experiment for which collimated 54 eV electrons were diffracted from the surface of a nickel crystal. The pattern measured was in exact agreement with that expected from x-rays. Independently in 1928, the English physicist George Thomson (1892–1975), also observed a diffraction pattern similar to that of x-rays, when sending a beam of 10 kilovolt electrons through a polycrystalline foil. The scattering angles measured agreed with those predicted by de Broglie's postulate.

George Thomson was the son of J. J. Thomson who discovered the electron. George demonstrated that an electron could be diffracted like a wave, providing additional evidence for de Broglie's particle-wave duality. In 1937, Clinton Davidsson and George Thomson shared the Nobel Prize in Physics "…for discovering the wave-like properties of the electron."

Electron Spin

In 1922, the German-American physicist Otto Stern (1888–1969) and the German physicist Walther Gerlach (1889–1979) conducted an experiment that was intended to disprove aspects of the Bohr atom. Instead, they discovered that the electron had an intrinsic angular momentum, like a miniature spinning gyroscope. Stern and Gerlach directed neutral silver atoms extracted from a hot oven that were sent through an in-homogenous magnetic field before impacting a metallic plate. Because the silver atoms had only one electron in their outer shell, the experiment permitted the magnetic properties of a single electron to be studied. The laws of classical physics predicted that the collection of condensed silver atoms on the plate should form a thin solid line in the same shape as the original beam. However, the in-homogeneous magnetic field caused the beam to split in two separate directions, creating two lines on the metallic plate.

In 1925, the Dutch-American physicists Samuel Goudsmit (1902–1978) and George Uhlenbeck (1900–1988) jointly explained this splitting into discrete spots by postulating that the electron had an intrinsic angular momentum (spin), which takes two possible quantized values (spin up or spin down), independent of its orbital characteristics. The idea arose earlier as an explanation for the fine-structure splitting observed in atomic spectra.

This celebrated Stern-Gerlach experiment was repeated in 1927 with hydrogen atoms, confirming the splitting, as with the silver atoms. It reinforced the interpretation of intrinsic electron spin. In 1943, Otto Stern was awarded the Nobel Prize in Physics for "…his contribution to the development of the molecular ray methods and his discovery of the magnetic moment of the proton" (ironically not for the Stern-Gerlach experiment). Historically, the Stern-Gerlach experiment was conclusive in establishing the correctness of angular-momentum quantization in atomic-scale systems.

Schrödinger and Heisenberg

Bohr's model was limited to one dimension and thus was applicable only to atoms with one electron. This model was unable to explain the presence of multiple orbitals and the fine structure of the spectra emitted in three dimensions from multiple electrons. In the meantime, de Broglie's concept of matter waves had been enthusiastically endorsed by Einstein, who saw matter waves as being an intuitive depiction of nature.

In 1925, hearing of de Broglie's concept, the Austrian theoretical physicist Erwin Schrödinger (1887–1961) was motivated to construct a three-dimensional mathematical model to describe the matter waves $\psi(x, t)$ in a potential field $V(x, t)$, similar to that of an atom. The equation he developed and published in 1926 took the form of a linear second-order partial differential equation that can be illustrated in one dimension as: $-\hbar^2/2md^2\psi(x, t)/dx^2 + V(x, t)\psi(x, t) = i\hbar d\psi(x, t)/dt$, where $\hbar$ represents Planck's constant divided by 2π. In three dimensions, the relation became the famous Schrödinger Wave Equation used throughout the fields of atomic, nuclear and solid-state physics. That equation attained the same importance to the field of quantum mechanics as the laws of Newton did for classical mechanics.

Schrödinger established the validity of his equation by calculating the spectra of the hydrogen atom, predicting many of its characteristics with great accuracy. The solution of the equation was the wave function amplitude ψ, which contains all the particulars that can be known about a system. The equation for ψ, was used to predict how particles move under the influence of a specified potential $V(x, t)$. Just a few days after Schrödinger's fourth and last paper was published, the German physicist Max Born (1882–1970) successfully interpreted ψ as a probability amplitude, whose modulus (absolute value) squared was equal to a probability density, i.e. that the modulus of ψ is the probability that, at a given moment of time, a hydrogen atom was in a particular spatial configuration.

In 1933, the Nobel Prize in Physics was awarded jointly to Erwin Schrödinger and Paul Dirac (1902–1984) "...for their discovery of new productive forms of atomic energy." In 1954, Max Born received the Nobel Prize in Physics for his "...fundamental research in quantum mechanics, especially in the statistical interpretation of the wave function."

As Schrödinger proposed his wave equation, the German physicist Werner Heisenberg (1901–1976) had already developed a matrix based model description of the atom. Although the two descriptions were mathematically equivalent, because of the unfamiliarity with matrix mechanics in the community, Schrödinger's approach prevailed. The matrix mechanics of Heisenberg, however, led him to the concept of the uncertainty principle, for which he became well known.

Heisenberg engaged in a *gedanken* experiment, in which he tried to find the position of an electron by measuring its location with a high-resolution microscope (i.e. one using short wavelength light). The light (photons) used to image the electron, gives the electron a nudge, changing its momentum in some random way. A better microscope (i.e. one using a more energetic photon, which had a shorter wavelength and thus greater resolving power), exacerbates the problem. A lower resolution scope provides a limited nudge but would be of poorer resolution. As the better microscope attempts to measure the location x of the electron (or particle) the more uncertain is its initial momentum p and vice versa. This line of reasoning led to Heisenberg's famous uncertainty principle, which stated: the uncertainty Δx in knowing a particles position x simultaneously with knowing its momentum p, with an accuracy of Δp, was limited by the relation $\Delta x \, \Delta p > h/2\pi$, where h was Planck's constant.

Previously, physicists thought that if one knew the exact position and momentum of a particle at any time and the forces acting on it, one could predict its exact position and momentum in the future. The uncertainty principle implied that there were limitations to this knowledge. Simultaneous determination of the position and momentum of a moving particle automatically contains errors. The product of the uncertainty of the position and momentum cannot be less than the Planck constant h divided by 2π. Although these errors are insignificant on the macroscopic scale, they cannot be ignored on the microscopic scale of the atom. A similar relation also holds for simultaneously knowing the time t a particle has a given energy E: that is $\Delta t \, \Delta E > h/2\pi$, where Δt and ΔE are the uncertainties simultaneously in knowing t and E, respectively. In 1932, the Prize in Physics was awarded to Heisenberg "...for the creation of quantum mechanics, the application of which, led to the discovery of the allotropic forms of molecular hydrogen."   .

The Neutron

Prior to 1932, physicists assumed that the nucleus was composed of A protons and (A–Z) electrons, where Z and A were defined as follows: the atomic number Z was the integer number of electrons in the given atom (i.e. the number revolving around the nucleus), the atomic weight A of an atom was the integer closest in weight to A times the mass of the hydrogen atom (or approximately A times the mass of a proton). Experimentally, A ~ 2 Z (except for hydrogen where A = Z = 1). These relations gave the correct mass and enabled the nuclear charge to be determined accurately. From the equations, however, it was clear that a nucleus with a mass of A could not be composed of A protons alone; the positive charge of the nucleus would be two times too large.

At the time, electrons and protons were the only two elementary particles known. The existence of electrons within the nucleus (needed to cancel the excess proton charge) gave rise to other issues. Years earlier, Rutherford had suggested the existence of a particle having the same mass as the proton, but zero charge (today called the neutron). Detecting such a particle was difficult because particle detection relied on observing the motion of particles having charge in electric or magnetic fields.

During the early 1930s, the French physical chemists Frédéric (1900–1958) and Irène (1897–1956) Joliot-Curie were studying the unidentified radiation generated downstream of a beryllium target impacted by energetic alpha particles from a radioactive polonium source. In their experiments, the unknown radiation was not influenced by electric fields and continued downstream to strike a paraffin wax target from which protons were measured on the downstream side of the target. They believed the unknown radiation to be gamma-rays (high-energy x-rays) that knocked the recoiling protons loose.

The English physicist James Chadwick (1891–1974), hearing of the Curies's experiment thought the explanation did not fit and repeated the experiments in England. His analyses suggested that gamma rays, having no mass, could not dislodge particles as heavy as a proton from a target. Being a disciple of Rutherford, Chadwick reasoned that the unknown radiation might be the elusive neutron. He began experiments demonstrating that the radiation was consistent with that of a neutron for which he had been searching.

In addition to the paraffin wax, Chadwick experimented with other targets, including lithium, nitrogen, and helium. These targets helped him estimate that the new particle had a mass of just slightly more than that of the proton. Because the neutrons had no charge, Chadwick observed they penetrated much deeper into a target than protons would. Convinced that it was a neutron that he was measuring and not a proton-electron pairing, in 1932 he submitted a paper with the title *The Existence of a Neutron*. Within a short time his results were accepted. After the electron and proton, the neutron became the third and final piece of the atom to be discovered. Eventually, the nucleus was determined to contain Z protons and definitively (A–Z) neutrons. The word nucleon became the term used for either the proton or neutron. In 1935, Chadwick was awarded the Nobel Prize in Physics "...for the discovery of the neutron." He was knighted in 1945.

The neutron became the key to the prodigious developments in nuclear physics that followed. The neutron made an ideal projectile for exploring the nucleus. Unlike like charged projectiles, it was not deflected by the Coulomb charge of the nucleus. Eventually, the neutron was used to impact the uranium atom and the world was changed forever.

Nuclear Discoveries

Nuclear Fission

With the neutron discovery in 1932, Chadwick's colleague, the English physicist Norman Feather (1904–1978) observed that neutron scattering from nitrogen (atomic number 7) induced nitrogen to disintegrate into boron (atomic number 4). In the process, an alpha particle was either ejected or the neutrons scattered and decayed, producing protons. Feather became the first to demonstrate that neutrons can produce the disintegration of a nucleus.

Neutrons soon became the diagnostic to probe nuclear structure. In Rome, the Italian physicist Enrico Fermi (1901–1954) studied the effects of irradiating heavier elements with neutrons. By 1934, he had used neutrons to induce radioactivity in as many as 22 different elements. By slowing the neutrons down in paraffin wax prior to neutron bombardment, Fermi determined that the radioactivity of the subsequently struck elements could be increased by over a hundred; and that the slow neutrons had a greater cross section (probability) for interaction than the fast neutrons. In 1938, Fermi was awarded the Nobel Prize in Physics "...for his demonstrations of the existence of new radioactive elements produced by neutron irradiation and for his related discovery of nuclear reactions brought about by slow neutrons."

In Berlin, between the years 1934 and 1938, the collaboration of Austrian physicist Lise Meitner (1878–1968), the German chemists Otto Hahn (1878–1968) and Fritz Strassmann (1902–1980) advanced the research begun by Fermi and his group when they struck uranium with neutrons. In late 1938, they did a pivotal experiment in which they concluded, with interpretation from the Austrian physicist Otto Frisch (1904–1979), that they had produced nuclear fission by splitting the uranium atom (atomic number 92) into much lighter nuclides (barium [atomic number 56], in particular).

The uranium nucleus was made unstable by the absorption of the neutron causing it to split into lighter elements that had greater stability. Their results were sent to *Nature* in January of 1939 and published February 11, 1939. Based on Einstein's relation $E = mc^2$, Meitner and Frisch estimated that each uranium atom that had fissioned, released about 200 MeV (1 eV = 1.6×10^{-19} J) of energy. Although a small amount, the amount of energy, when converted to macroscopic units of energy emitted per pound of uranium, the energy released from the fission of uranium was approximately a million times larger than the energy emitted per pound from burning coal.

Physicists soon realized that if a fission reaction could emit enough secondary neutrons, a chain reaction would likely occur, potentially releasing vast amounts of energy. In 1944, Hahn was awarded the Nobel Prize in Chemistry "...for his discovery of the fission of heavy atomic nuclei." Hahn is considered the father of nuclear chemistry.

Nuclear Chain Reaction

In 1938, Fermi emigrated to America, initially to Columbia University where he duplicated many of his earlier experiments. On January 25, 1939, in the basement of Columbia's Pupin Hall, Fermi and his team conducted the first nuclear fission experiments in America.

As with Fermi's move to America from Italy, the Hungarian physicist Leo Szilard (1898–1964) left Europe for the United States. Both were directly affected by the Nazi discrimination of Jews and foresaw the coming outbreak of war. Eventually, Szilard also found his way to Columbia University. There, upon hearing that nuclear fission had been induced in uranium by the Germans, he realized that uranium might be capable of sustaining a nuclear chain reaction. At Columbia, he tried to interest Fermi in such an experiment, using neutrons and uranium.

Earlier Szilard had been thinking about this concept of a neutron-induced chain reaction (though not necessarily with uranium). In 1933, he had filed for a patent on this concept in the United Kingdom. Unable to convince Fermi; Szilard was able to convince others for funding. With permission from the head of Columbia's Physics Department, Szilard designed an experiment on the seventh floor of Pupin Hall to irradiate natural uranium with neutrons from a radium beryllium source. There, he and his colleagues observed significant neutron multiplication, demonstrating that the fission of uranium produced more neutrons than it used, suggesting that a chain reaction was possible.

Subsequently, Szilard, Fermi and others grasped the military application of this enormous power. Knowing of the German research, they quickly drafted a letter, warning President Roosevelt (1882–1945) of the danger of a potential German nuclear bomb. Einstein, a colleague of Szilard, had left Germany in 1933, becoming a Professor of Theoretical Physics at Princeton University and then an American citizen in 1940. Einstein had achieved the necessary prestige and also signed the letter, which was delivered to the president on October 11, 1939. The letter led first to establishing research into nuclear fission by the US (United States) government and later to the birth of the secret American program to create a nuclear bomb, code named the Manhattan Project. The fear that Nazi Germany might develop a nuclear weapon motivated the US to achieve that goal first.

With funding and design help from Szilard, Fermi moved his team to the University of Chicago, where he assembled the first nuclear pile (a reactor), in a university squash court in November 1942. The nuclear pile was composed of 45,000 ultra-pure graphite blocks weighing 330 tons that acted as a neutron moderator. It was powered by 46 tons of natural uranium and uranium oxide. On December 2, 1942, the nuclear pile went critical, becoming the first self-sustaining nuclear chain-reaction made by humans. With this demonstration, the nuclear pile became the first technical step in the Manhattan Project, signaling the beginning of the Nuclear Age.

Nuclear (Atom) Bomb

During the next three years, a large scale research effort went into developing the necessary fissionable material suitable for a small nuclear bomb. The project connected three principal laboratories in secret locations across the country: the University of Chicago, Illinois; Oak Ridge, Tennessee; and the newly established headquarters of the project at Los Alamos, New Mexico. More than 6000 scientists were recruited in the effort. In 1943, the American theoretical physicist J. Robert Oppenheimer (1904–1967) became the director of the new laboratory at Los Alamos. He guided its construction and led the associated science program. At Los Alamos, Oppenheimer gathered together many of the great minds in physics, including some of the English nuclear pioneers, like Chadwick. Many of the scientists were Jewish and had escaped Europe earlier while they still could. Because of his successful leadership at Los Alamos from 1943 through 1945, Oppenheimer is often referred to as the father of the atomic bomb.

During the Manhattan Project, techniques for isolating the more fissionable isotope of uranium $^{92}U^{235}$ from the more abundant isotope $^{92}U^{238}$ evolved. (An isotope is two or more forms of the same element but having differing numbers of neutrons; the superscript on the left refers to the atomic number; that on the right the atomic weight.) Separating the necessary $^{92}U^{235}$ from $^{92}U^{238}$, however, was costly; the $^{92}U^{235}$ accounts for only 0.7% of the natural uranium. One workable process involved a lengthy procedure using centrifuges, where the given isotope was selected based solely on the weight difference.

The high cost of this procedure prompted a search for other fissionable materials. One byproduct of the Manhattan Project was the development of a nuclear reactor. Indirectly, it became immediately useful for the production of fissionable pluto-nium $^{94}Pu^{239}$. Very little plutonium occurs naturally in nature, however, plutonium is the result of uranium absorbing neutrons in a nuclear reactor. Because plutonium's atomic number was different from uranium it was easily separable from the uranium chemically. As such, it became the material of choice for the bomb. On July 16, 1945, the use of plutonium was successfully demonstrated at the Trinity nuclear test site in New Mexico.

Although Germany surrendered on May 7, 1945, World War II with Japan was still continued with great loss of life. President Harry Truman (1884–1972) was advised that any attempt to invade Japan would result in even greater loss of American life. Accordingly, Truman ordered that the new bomb be used to bring the remaining war to a quick end. On August 5, 1945, the first nuclear bomb, code named Little Boy, made of 64 kg $^{92}U^{235}$ was dropped on the Japanese City of Hiroshima. It exploded with an energy of 15 kilotons of TNT, instantly killing 100,000 people and perhaps another 100,000 later from radiation poisoning. It may not have been since the loss of 80,000 Roman solders by Hannibal at Cannae, Italy in 216 BCE that so many lost their life in a single day.

A second bomb, code named Fat Man, dropped on Nagasaki four days later on August 9, 1945, solicited an unconditional surrender from the Japanese. In contrast to Little Boy, which used uranium, Fat Man used 6.4 kg of plutonium. It exploded with a blast equivalent to that of 21 kilotons of TNT. The result was the immediate deaths of 40,000–75,000 people.

The example of the devastation wrought on Japan by the two nuclear explosions put an end to further use of nuclear weapons in World War II (1939–1945) and to war in general among the major world powers. Weapon development, however, did not cease with World War II. A Cold War (1947–1991) between the West and the communist Soviet Union (1922–1991) ensued when it advanced into Eastern Europe, seizing control, and forming puppet states in Poland, Hungary, Czechoslovakia, East Germany and the Baltic states.

Thermonuclear (H) Bomb

The massive destruction of the targets caused by the two nuclear bombs in Hiroshima and Nagasaki illustrated the large scale release of energy from the conversion of mass to energy through Einstein's relation $E = mc^2$. The energy released was a consequence of nuclear fission, in which a large nucleus splits into two intermediate size nuclei. This process occurs because the final nuclear state was more stable than the initial state. The binding energy of a nucleon (proton or neutron) to a large nucleus is 7.5 MeV compared to 8.5 MeV, the nucleon binding energy characteristic of that in either of the two resulting intermediate size nuclei. The energy liberated by fission, thus becomes approximately 1 MeV per nucleon.

In contrast to this process of nuclear fission, the process of nuclear fusion is that of combining two or more smaller and less stable nuclei to form a larger, more stable nucleus. For small nuclei, the trend in binding energy with nuclear size is reversed and the binding energy increases. Under extreme temperatures of millions of degrees the positively charged nuclei can gain sufficient kinetic energy to overcome their mutual Coulomb repulsion. When close enough, the short-range strong nuclear force overwhelms the electrostatic repulsive force and nuclei merge, as in compressing isotopes of hydrogen (deuterium $^1D^2$ or tritium $^1T^3$) together to form the stable helium atom ($^2He^4$) or as happens with hydrogen in the Sun, where helium forms under the pressure of gravity. In all cases, nuclear fission or fusion, the energy released is based on the fundamental point that the mass of a resultant nucleus or nuclei is less than its constituent parts.

During 1943, the Hungarian theoretical physicist Edward Teller (1908–2003) was invited to join the secret Manhattan Project. He left Hungary in 1933, eventually becoming an American citizen in 1941. During his time at Los Alamos, he worked on the idea of a super bomb where the isotopes of hydrogen would combine under extremely high temperatures and pressure to form helium in the process of nuclear fusion. Many of Teller's colleagues, i.e. Oppenheimer and Fermi, were confident the project was technically unfeasible. By the wars end, none of the designs had shown merit.

Nevertheless, Teller pursued the idea of a super bomb. Together with a group of scientists he continued direct approaches to the Joint Committee on Atomic Energy and to the military for its potential. On August 29, 1949, the Soviet Union unexpectedly tested a nuclear bomb. A short time later, the American and British intelligence came to the conclusion that the German born theoretical physicist Klaus Fuchs (1911–1988), was a spy passing information to the Soviets. He had been an integral part of the Manhattan Project. This event, together with the expansion of the Soviet Union into East Europe, led President Truman on January 31, 1950 to approve extensive funding for America to achieve the super bomb before the Soviets could.

In 1950, Teller returned to Los Alamos to work on the project. Together with another of Teller's former Manhattan Project colleagues, the Hungarian mathematician Stanislaw Ulam (1909–1984), the breakthrough came in 1951. The Teller-Ulam concept was to separate the fission and fusion components. The x-rays produced by a fission bomb (the primary) would be reflected off a surrounding radiation case to compress the fusion fuel (the secondary) before igniting it. With this configuration, there would be no limit on the energy yield from such a weapon.

The first detonation to evaluate the concept, code named Ivy Mike, was on November 1, 1952, at an Island currently one of the Marshall Islands in the Pacific. The test developed the equivalent explosive energy of 10 megatons of TNT, about 500 times the energy of the bomb dropped over Nagasaki. The island was nearly destroyed, becoming partially submerged under water. With its success, Teller became known as the father of the hydrogen bomb.

In 1952, Teller co-founded Lawrence Livermore National Laboratory with the American physicist Ernest Lawrence (1901–1958). The laboratory was developed to provide competition with Los Alamos and to stimulate innovation in nuclear energy. Teller became both its director and associate director for many years. Over time, he was honored with various prestigious awards including: the Albert Einstein Award in 1958, the Enrico Fermi Award in 1962 and the Presidential Medal of Freedom Award in 2003.

On August 12, 1953, the Soviet Union exploded the first deliverable H-bomb (hydrogen-bomb) with a yield of 400 kilotons of TNT, designed by the Russian physicist Andrei Sakharov (1921–1989). Nearly a year later, the Americans detonated its first practical H-bomb, which yielded 15 megatons. The nuclear arms race began.

In 1957, the English detonated their first thermonuclear device, followed by China in 1967, France in 1968, and both India and Pakistan in 1998. In 1963, a PTBT (Partial Test Ban Treaty) was signed by the major world powers, which banned nuclear-weapon tests in the atmosphere, outer space and underwater. The PTBT has been succeeded by the CNTBT (Comprehensive Nuclear-Test-Ban Treaty) in 1996, which prohibits "...any nuclear weapon test explosion or any other nuclear explosion" anywhere in the world. New signatures have been added to the treaty over time. To date there have been no violations.

Nuclear Beta Decay and the Neutrino

At the end of the nineteenth century, Rutherford identified that radioactive nuclei decay in three ways, which were labeled with the first three letters of the Greek alphabet: alpha particles, that were known to be ionized helium nuclei, beta particles, that were electrons and gamma rays that were high energy photons. These decays were a mechanism by which a nucleus could become more stable. By the early 1930s, measurements of the alpha particles and gamma rays appeared to conserve energy in their decay process. Careful measurements of the beta radiation, however, suggested otherwise. The observed energies of the electron and recoiling nucleus (containing the residual proton from the neutron decay within the nucleus into the electron) were not consistent with the laws of conservation of momentum and energy.

Rather than give up on these conservation principles, the Austrian theoretical physicist Wolfgang Pauli (1900–1958), proposed that there was a third particle involved in beta decay, in addition to the electron and nucleus. The third particle would have no charge and little, if any mass, and would simply carry off the missing energy and momentum in keeping with conservation principles. In beta decay, a neutron in the nucleus would decay into a proton, an electron and the third particle. In 1932, Fermi named it a neutrino to distinguish it from the recently discovered neutron by Chadwick. This new process was referred to as nuclear beta decay.

In the same year, the American physicist Carl Anderson (1905–1991) discovered a positron by the tracks left by cosmic rays in a cloud chamber. The positron was the antiparticle of the electron, being identical to the electron, except for being of positive charge. The positron had been predicted in 1928 by the English theoretical physicist Paul Dirac. He demonstrated mathematically that Einstein's STR implied that every particle has a corresponding antiparticle, each with the opposite charge, but with the same mass. For his work, Dirac was awarded the Nobel Prize jointly with Schrödinger, in 1933. Anderson received the Nobel Prize for his discovery in 1936.

During this same period, the German theoretical physicists, Hans Bethe (1906–2005) and Rudolf Peierls (1907–1995), concluded that the inverse of beta decay ought be theoretically possible. Such a decay might involve a neutrino interaction with a proton (in a nucleus) producing a neutron and positron. In 1934, they estimated that the neutrino proton scattering cross section would be very low and that the neutrino would traverse the entire Earth without interacting. Therefore, in 1934, searching for such a particle was not feasible.

The nuclear fission reactors designed to produce plutonium for the war effort, however, generated large fluxes of neutrinos, *e.g.*, 5×10^{13} neutrinos per second per square centimeter. With World War II over, it was an opportune time for making a concerted effort to look for this elusive neutrino. The American physicists Frederick Reines (1918–1998) and Clyde Cowan (1919–1974), working at Los Alamos Laboratory, set up a 10-ton detector system to look for the neutrino, using its predicted inverse beta decay reaction.

The neutrino from a reactor would impact a proton in a 200 L tank of water and produce a neutron and a positron. Cadmium chloride salts dissolved in the water were a highly effective neutron absorber that generate a gamma ray several microseconds later when it absorbed a neutron. The neutrino was uniquely identified by the observation of two 0.51 MeV gamma rays from a positron annihilation with an electron, followed by the gamma ray from the decay of the cadmium.

The experiment by Reines and Cowan at the Savannah River Reactor in South Carolina, which had excellent shielding against cosmic rays, generated a measured neutrino cross section of 6.3×10^{-44} cm^2, in excellent agreement with the 6×10^{-44} cm^2 predicted. No events were detected when the reactor was shut down. Their results were published in *Science* on July 20, 1956. Reines was honored with a share of the Nobel Prize in Physics in 1995 for his work with Cowan, being the first to detect the neutrino.

The area of 6.3×10^{-44} cm^2 is extremely small. The typical unit of area used in nuclear physics reactions is that of a barn. The barn is the approximate cross sectional area of a uranium nucleus and is defined to be 10^{-24} cm^2. The area of interaction for the measured neutrino was thus about 10^{20} times smaller than a barn, underscoring its very small probability of interacting with matter and the difficulty of its being detected.

Nuclear Sun and Solar Neutrinos

The origin of the Sun's energy and its age, had long been a mystery. The first real supposition came indirectly from the English chemist and Nobel Laureate Francis Aston (1877–1945). He had made precise measurements of masses of different atoms, including hydrogen, in the 1920s. Of relevance to the conundrum was his observation that the mass of four hydrogen nuclei (the nuclear ingredients of helium) was slightly heavier than a helium nucleus. The importance of this difference in mass was immediately appreciated by Sir Author Eddington. He theorized that if hydrogen were to be converted into helium, the Sun could shine for a very long time, approximately 100 billion years, using Einstein's relation converting mass into energy $E = mc^2$.

By 1938, Hans Bethe, the recognized authority on nuclear reactions, had evaluated the basic nuclear processes by which hydrogen is fused into helium in the interiors of stars. These two elements are among the most abundant elements in the universe.

Bethe submitted the results of his analyses in a paper titled "Energy production in Stars", to *Physics Review*, which was published in 1939. One of the many processes that he had analyzed, was the sequential double proton-proton chain that generated helium out of hydrogen. It was this reaction that was calculated to be the dominate energy source in the Sun and in stars like or less massive than the Sun. The processes are expressed as four hydrogen nuclei evolving into helium, plus two positrons, plus two neutrinos, plus energy. The Sun's extensive gravitational energy compresses (fuses) the heavier hydrogen nuclei into the lighter helium nuclei producing energy, as in the hydrogen bomb.

The interiors of the Sun and stars, were controlled thermonuclear explosions on a vast scale. Bethe's theory, was the first to provide a basic understanding of how stars radiate and evolve with time. His theory led to the successful evaluation of the measured luminosities of the Sun and similar stars. The concept that stars are powered by nuclear fusion became one of the foundations of modern astronomy.

Bethe, like many of the other Jewish scientists, left Germany in 1933 and went first to England and then to America. In 1935, he accepted a position at Cornell University and, in 1941, became a naturalized citizen. By 1943, Oppenheimer, recognizing his talents, appointed him director of the Theoretical Division at Los Alamos. During his tenure, he contributed significantly to not only the Manhattan Project but also to later developments of thermonuclear fusion. Bethe was a strong proponent against nuclear testing and the nuclear arms race and continued to make significant contributions to science. In 1967, Bethe received the Nobel Prize in Physics for his work on the theory of stellar nucleosynthesis.

Theoretically, the Sun generates 2×10^{38} neutrinos per second, constituting 3% of its radiated energy. These weakly interacting neutrinos bathe the Earth's surface with 10^{11} neutrinos per cm^2 per second. In 1964, the American chemist and physicist Raymond Davis Jr. (1914–2006) estimated that these solar neutrinos could realistically be searched for in the laboratory, despite their extremely weak interaction probability, helping to confirm the fusion mechanisms of the Sun's interior. Davis surmised that a solar neutrino would produce radioactive argon (^{18}Ar) when it encounters a chlorine nucleus (^{17}Cl) and that the reaction would be observable in perchloroethylene, a utility dry cleaning fluid.

In 1968, Davis developed a very large detector based on this concept. He placed a 100,000 gallon tank of perchloroethylene, 1.46 km underground in the Homestake Gold Mine of South Dakota to shield it from background cosmic rays. Davis's experiment confirmed that the Sun produced fusion neutrinos. The number of neutrinos measured, however, was only one third of that predicted. It took more than 20 years and the evolution of the field of particle physics to understand this deficit. In 2002, Davis shared the Nobel Prize in Physics for being the first to detect neutrinos emitted from the Sun.

Quantum Electrodynamics (QED)

While efforts to understand atomic and nuclear physics during the first half of the twentieth century were underway, a related effort resulted in discoveries of the wave nature of electrons, particles and constraints imposed by Einstein's STR. Standard quantum theory as developed by Bohr, Schrödinger and Heisenberg in the 1920s was adequate for describing the workings of individual particles in isolation and at non-relativistic speeds. But to explain their interactions in general, physicists required further research.

The new theoretical modeling needed, was first explored in 1928 by the English physicist Paul Dirac (1902–1984). He developed a mathematical model that included quantum mechanics and the STR and that took account of both the electron spin and motion. It was a relativistic theory because Einstein's theory was built into its equations. Dirac referred to the new theory as quantum electrodynamics (QED). As previously mentioned, he shared the 1933 Nobel Prize in Physics with Schrödinger. Subsequently, the theory was advanced by many theoretical physicists, including Pauli, Heisenberg, Fermi and Bethe.

By the late 1940s, QED was clarified by the American theoretical physicists Richard Feynman (1918–1988) and Julian Schwinger (1918–1994) and the Japanese theoretical physicist Sin-Itiro Tomonaga (1906–1979). This theory was based on the concept that charged particles interact by emitting and absorbing photons, the particles that propagate the electromagnetic force. The emission and absorption processes are virtual, in that the photons cannot be seen or detected in anyway. These actions became possible because they can do so in accordance with Heisenberg's uncertainty relations, even though their presence violates conservation of momentum and energy. Heisenberg's relations imply that both the position and speed of a particle, i.e. a photon or electron, cannot be known simultaneously with perfect accuracy. The Heisenberg mechanism allows the particles to borrow energy or momentum from the vacuum for a finite amount of time.

The photon exchange becomes the force of the interaction, when the interacting particles change their speed and direction of travel as they release or absorb the energy of the photon. Alternatively, the photons can be emitted in a free state, in which situation they may be observed as light or as other forms of electromagnetic radiation like x-rays or radio waves. In addition, QED claims the vacuum between interacting particles is not just empty space. Energy in the electromagnetic field can momentarily transform into matter and antimatter pairs like an electron and its antiparticle, the positron. These short-lived virtual pairs, referred to as vacuum bubbles, polarize the vacuum. The pairs disappear almost immediately but during their limited existence they slightly alter the shape of the electric field. The electron–positron pairs aligned themselves along the field lines, partially counteracting the original field (a screening effect, as in a dielectric between the plates of a capacitor). The field, therefore, becomes weaker than would be expected if the vacuum were completely empty.

In an atom, the random electric fields attracting a negative electron to the positive nucleus would be infinitesimally reduced; more so if the orbit is closer to the strong field of the nucleus, *i.e.,* more tightly bound to the nucleus. In the spectral line emission process, the photon energy would be slightly less, leading to a subtle shift in the energy of any transition photon.

In 1938, the American physicist Willis Lamb (1913–2008) joined the Columbia University physics faculty where he worked in Columbia's Radiation Laboratory during World War II. He developed ways to generate microwaves of higher frequencies and energies for use in radar. After the war, he envisaged the same technology could be used to precisely examine the hydrogen line emission spectrum calculated earlier by Dirac. Dirac had used his mathematical model based on the classical vacuum to predict the lines that appear in hydrogen's emission spectrum.

In 1946, Lamb applied his new methods, with its greater resolution, to measure the lines of hydrogen. In 1947, he discovered that their frequencies were slightly different from what was expected. Lamb found two of Dirac's lines (whose associated electron was in an orbit closest to the nucleus) predicted identical energies that were not identical. Lamb's precise measurements demonstrated they differed by approximately one part in a million. This difference was subsequently shown to be a result of the interaction between fluctuations of the vacuum energy and the hydrogen electron in those close orbitals. It was referred to as the Lamb Shift, due to the subtle effects of QED, which explained the lines exactly.

Lamb's work became one of the foundations of quantum electrodynamics. He was the first to show that the void was not empty but could be a cauldron of virtual particles of matter and antimatter in the presence of electric fields. In 1955, he received the Nobel Prize in Physics for his experimental work on the fine structure of the hydrogen atom and for the discovery of the Lamb Shift.

Ten years later, in 1965, the Nobel Prize in Physics was awarded to Richard Feynman, Julian Schwinger, and Sin-Itiro Tomonaga "…for their fundamental work in quantum electrodynamics, with deep-ploughing consequences for the physics of elementary particles." Because of QED's extremely accurate predictions of the electromagnetic behavior of atoms, Feynman named the theory "…the jewel of physics." QED became the model for the subsequent quantum field theories of the new particles being discovered in the next half century.

Chapter 13
Particle Physics: 1950–2023

Cosmic Rays

By the 1950's, electrons, protons and neutrons were thought to be the fundamental building blocks of matter. Measurements of cosmic rays in cloud chambers or photographic emulsion plates, during the 1930s and 1940s, however, suggested the existence of other particles as well. In 1936, the American physicists Carl Anderson (1905–1991) and Seth Neddermeyer (1907–1988) found a weakly interacting particle that they named the muon, in their cloud chamber. In the chamber, particles were identified by vapor trails that were left behind their passage in the apparatus. In an other kind of cosmic-ray detector, photographic emulsions plates were often positioned at high-altitude mountain locations for long periods of time. Following the development of the plates, microscopic viewing of the emulsion often revealed tracks of charged particles.

By 1947, strongly interacting particles called pions were identified in such experiments by the English physicist Cecil Powell (1903–1969) and his colleagues. In the emulsions, a pion was sometimes observed to decay into a muon, which decayed into an electron. The masses of the charged pion and muon were eventually determined to be 140 MeV/c^2 and 106 MeV/c^2, respectively. The latter is about 200 times the mass of the electron, which is 0.511 MeV/c^2 and both are less than the mass of any atom. (For reference, in particle physics, because $E = mc^2$, the common unit of mass used is the MeV/c^2, which is directly related to energy. One eV is the energy gained when an electron undergoes a potential difference of one volt. One MeV is one million eV: 1,000,000 eV.)

In 1935, the Japanese theoretical physicist Hideki Yukawa (1907–1981) reasoned that just as the electromagnetic force was transmitted by the photon (a particle), the attractive strong force that holds the protons together against their repulsive positive charge in the nucleus could also be carried by a particle.

Yukawa theorized that the more massive the particle the shorter would be its range and that the mass of this particle likely would lie between that of an electron and proton. Yukawa's particle was named a meson (later a pion or pi meson), from the Greek for in-between. Initially the muon, was identified as Yukawa's hypothetical particle, as it existed within the expected mass range. But, with its weak interaction with matter and the pion's strong interaction, the names were later reversed. These particles were among the first to be identified outside those thought to form the basic constituents of matter. In 1949, Yukawa was awarded the Nobel Prize in Physics for his theoretical prediction of the existence of mesons. Powell received the Nobel Prize in Physics the following year for the development of the photographic emulsion method of studying nuclear processes and for the resulting discovery of the pion.

After World War II, there was a significant increase in the support of physics, in particular, facilities to generate and study the cosmic particles. In America, government funding for understanding the new particles was combined with nuclear physics. The Berkeley Radiation Laboratory and the Brookhaven National Laboratory were entirely supported by the Atomic Energy Commission, for example; and the Columbia University's Nevis Synchrotron Laboratory was supported by the office of Naval Research,.

Between 1946 and 1952, 10 synchro-cyclotrons were built in America. Some machines, accelerated protons, like that at the Nevis Laboratory, and others accelerated electrons. Their impact with targets generated particles like the pion, if the energy was high enough. As the machine energy increased, more massive particles were produced. In addition, several facilities developed in Europe. One of the early laboratories was referred to as CERN (French: *Conseil European pour la Researche Nucleaire*) located just west of Geneva, Switzerland and adjacent to the Swiss-French boarder. CERN was founded in 1954 to bring pure science back to post war Europe and to recover the European physicists who had emigrated to America as a result of World War II. It was dedicated to the peaceful understanding of the world's smallest objects and built to bring insights toward nature's most basic building blocks of the universe.

The construction of high-energy particle accelerators and their variations formed the foundation of the evolving field of HEPP (high-energy particle physics), which motivated the search for unexpected, as well as, predicted particles. Concurrent with accelerator development was the associated improvement in particle detection. With the complexity of the experiments, the number of physicists on the published papers expanded exponentially from a few in 1950 to hundreds by 1990. To avoid disagreements, the ordering of the authors was made alphabetical. The fifty years between 1950 and 2000 formed the golden years of HEPP, with new discoveries occurring continuously throughout each of the decades.

Many of these discoveries, including new types of particle accelerators and diagnostics, established their importance in the medical field. More than 10,000 linear electron-accelerators are being used worldwide in radiotherapy. Similar to x-rays in the last century, PET (positron-emission tomography) and NMR (nuclear magnetic resonance) more commonly referred to as MRI (magnetic resonance imaging), have become standard medical diagnostics. To enable communication among the expanding groups within the developing collaborations, the WWW (World Wide Web) was first developed at CERN.

The historical development of the field of HEPP highlighted here, by a few of the influential scientists and their experimental discoveries, has lead to the SM (Standard Model of Particle Physics). It is this theoretical model, backed by experiment, that currently forms humanities basic understanding of matter and radiation, together with the forces that hold the universe together. This knowledge provides the comprehension of the immediate aftermath of the Big Bang and which has lead to the rational search for a Higgs boson.

Symmetry Violations and Family-Concept

Pions and Muons

One of the early experimenters in the new field of particle physics was the German-American physicist Jack Steinberger (1921–2020). In 1947, his Ph.D., motivated by Enrico Fermi, was to investigate cosmic muon decay by measuring the energy spectrum of its resulting decay electron. From these measurements, Steinberger inferred muon decay was accompanied by two neutral neutrinos. He used 80 thin-walled Geiger counter tubes sandwiched between absorbers to measure the range of the electrons at the top of the 4000 m Mount Evens in Colorado, where the muon flux was three times that at sea level. His experiment was challenging at the time. Partly as a result of the experiment's success, Steinberger earned a Ph.D. from the University of Chicago in 1948.

In 1950, the Nevis cyclotron at Columbia University, having a proton-beam energy of 400 MeV, became an efficient source of pions. Once the protons accelerated to maximum energy in the cyclotron, they collided with a target located near the exterior of the interior edge of the cyclotron. There, pions were produced, which exited through a channel in a 2 m thick radiation shielding wall that surrounding the accelerator. At Nevis, Steinberger explored the fundamental properties of the pion, including its lifetime, spin, parity and other of its quantum numbers. He also observed its interaction with protons and neutrons.

A result of such experiments was the observation of the 3/2 resonance in pion-nucleon scattering. It was the first of many examples of hadronic resonances discovered. Subsequently, hadrons became defined as the strongly interacting particles like the nucleons (protons and neutrons) and pions, while leptons were defined as the weakly interacting particles like the electron, muon and neutrino.

Parity Violation

Until the middle of the twentieth century, spatial symmetries were assumed to be a major feature of nature. The concept that the laws of physics, whether viewed in a mirror reflection or not, were thought to be identical. This law of parity symmetry (reflection symmetry or spatial inversion) was thought to be true for all interactions in nature. The concept subscribed to the idea that the current world and one built like its mirror image behaved in the same way, with the only difference being that left and right would be reversed.

Parity conservation had been experimentally confirmed in both electromagnetic and the strong interaction of nuclear forces. In 1956, to explain certain particle decays in the newly found K-mesons (kaons), the Chinese-American theoretical physicists Tsung-Dao Lee (1926–) and Chen-Ning Yang (1922–) proposed that parity symmetry in weak decays might be violated. They suggested to the Chinese-American experimental physicist Chien-Shiung Wu (1912–1997) that she perform an experiment to measure the direction of beta-decay electrons from the weak decay of the cobalt nuclei relative to its direction of polarization. Wu was an expert on beta-decay spectroscopy. Beta-decay in a cobalt nucleus occurs when one of its neutrons decays via the weak interaction into a proton forming a nickel nucleus. In this reaction, the neutron emits an electron e and neutrino v. The excited nickel nucleus subsequently gives off two gamma rays γ. Summarizing, the $^{27}Co^{60}$ evolves to $^{28}Ni^{60} + e + v + 2\gamma$.

The purpose of Wu's experiment was to determine whether the decay products of cobalt-60 were being emitted preferentially in one direction, or not, relative to its polarization (axis of spin). If there was a preferential direction, this would signify the violation of parity symmetry; if the weak interaction were parity conserving, the decay emissions would be emitted with equal probability in all directions.

In 1956, Wu conduced this difficult cryogenic experiment at the National Bureau of Standards where the spin of the cobalt was carefully controlled. Her earliest experiments indicated that the emission of electrons with respect to the spin of the cobalt (and the simultaneous gamma-ray emission via the electromagnetic process) was asymmetrical.

The American experimental physicist Leon Lederman (1922–2018), a close colleague of Lee and Wu at Columbia, heard of the large parity violation of Wu early in 1957. Lederman thought if parity violation was so large, he might be able to observe it as in weak muon decay into electrons at the Nevis cyclotron. There, the muons were produced at the end of the long decay process in which pions first decayed weakly into muons. Confirmation of this result was needed.

Lederman was an expert in extracting pions from Nevis (being the first to so, using its magnetic fringing field). Within a short time, Lederman devised an experiment, together with Richard Garwin (1928-) a colleague from IBM, to measure the direction of electrons from the decay of polarized spinning muons, stopping in a carbon target. The target was imbedded within an oscillating magnetic field, which changed the orientation of the spinning muons. After a few days of data accumulation, they observed a large asymmetry in the direction of the electrons with respect to the muon's spin direction, definitively confirming parity violation in weak decays. Lederman delayed publication of their results until Wu had confirmed her results, and the two papers, appeared back-to-back in the same *Physical Review* journal on February

15, 1957. This unexpected asymmetry in nature became the first of several to be discovered in particle decays.

In 1957, Lee and Yang were awarded the Nobel Prize in Physics for their work on parity violation in weak interactions that Wu experimentally verified in 1956 and was confirmed by Lederman. Wu was later awarded the first Wolf Prize in physics in 1978 for her work. Lederman became the Director of Nevis Laboratory in 1961, remaining as Director through 1978. The Wolf Prizes were established by the Wolf Foundation in Israel for excellence in agriculture, chemistry, mathematics, medicine, physics and the arts. They are considered second in prestige after the Nobel Prize.

Lederman's parity experiment and Steinberger's experiments at Nevis, became the first of many seminal discoveries they both made during their long lives. These two experimental physicists, Lederman and Steinberger, were integral to the evolution of particle physics from the beginning, which led 50 years later to the Standard Model of Particle Physics.

Two Different Neutrinos

The prevailing theory of weak interactions (originally due to Fermi) suggested that when the muon decays into an electron and two neutrinos, the pair should have a finite probability to merge, creating a photon. However, no photon had ever been measured. This failure suggested that the neutrino and its partner neutrino were of two different kinds that could not combine. Moreover, knowledge of any difference in the neutrinos might provide some insight into why the electron and muon were identical in every respect, aside from the muon being 200 times heavier.

To examine this issue, intense neutrino beams were required. By 1961, Brookhaven National Laboratory's AGS (Alternating Gradient Synchrotron), located on Long Island, New York, had just come online to produce the world's most energetic 30 GeV (1 GeV = 1 × 10^9 eV) proton beam. The 257 m diameter AGS ring housed 240 magnets in which the protons were contained magnetically, accelerated electrically and directed into a beryllium target, producing large numbers of pions. Taking advantage of the new capability, in 1961, Leon Lederman, Melvin Schwartz (1932–2006) and Jack Steinberger, together with the Brookhaven staff and others, designed a neutrino-beam experiment to determine if the neutrinos from the decay of the pions into muons and neutrinos were different from those measured in nuclear beta decay. (Schwartz was a student of Steinberger and later became a professor at Columbia University.)

The energy of the protons was reduced to 15 GeV to limit the amount of shielding required for the experiment. An aperture in the accelerator's shielding wall enabled an intense collimated beam of pions, which peaked at 3 GeV, to escape. The pions, together with their decays, formed a 14 degree radiation cone of neutrinos over a flight path of 20 m before being stopped by a shielding wall. Over that distance, 10% of the pions decayed into muons and associated neutrinos.

The Columbia University experiment, designed to measure these neutrinos, was set up behind the wall, which was a 2000 ton, 13.5 m thick, steel shield. The pions, which interact through the strong nuclear force that operates inside the nucleus, were stopped within a distance of 1/2 m. The muons, which interact only through the electromagnetic force and weak interaction, loose their energy by colliding with the atomic electrons, penetrated deeper. The thickness of the steel shield, which was made from recently decommissioned battleships, was designed to stop these muons. In contrast, most neutrinos would just pass through the shield. To isolate the experiment from cosmic ray contamination, the experiment was further shielded with 5.5 m of concrete on the ceiling and floor of the detector.

The neutrinos were detected in a 10 ton spark-chamber system consisting of ninety 2.52 cm thick vertical aluminum plates separated by gaps filled with neon gas. Occasionally, a neutrino would strike a proton within the aluminum and the effect of the collision was detected, using an enclosed scintillation counter, which triggered a high voltage to the chambers. The voltage initiated a visible spark to form between the gaps along the path of a forward moving, gas-ionizing particle produced in the collision. A picture of the sequential sparks, when viewed ninety degrees to the sparks, provided an image of the track of the triggering particle passing through the chambers.

Throughout this experiment, which ran intermittently from September 1961 to July 1962, an estimated 3.5×10^{17} protons impacted the beryllium target. As a result, 10^{14} neutrinos passed through the detector with 51 particles detected in the spark chambers that were consistent with a beam generated neutrino. Only muons were measured within the detector. They were explained by the reaction: neutrino + proton turns into neutron + muon). No electrons were measured. This *tour de force* experiment, the largest in the world at the time, concluded that neutrinos from pion decay (where the pion decays into a muon and neutrino), were different from the neutrinos involved in electron beta decay (where the neutron in a nucleus decays into a proton, electron and neutrino). If they were the same, an electron would have been expected instead of the measured muon. In 1962, the results were published in *Phys. Rev. Letters*, the premier physics journal at the time.

This experiment became the beginning of beta decay neutrinos, with their associated electron, being referred to as electron neutrinos. Neutrinos arising from pion decay, with their associated muon, were now named muon neutrinos. Similar to a grouping that paired the electron neutrino with an electron, this experiment established a pairing between the muon neutrino with a muon. The discovery of the muon neutrino eventually led to the recognition of a number of different families of subatomic particles. This observation contributed to the Standard Model's classification, used to order all known elementary particles. In 1988, Lederman, Schwartz and Steinberger were recognized for its importance and were awarded the Nobel Prize in Physics "...for the neutrino beam method and the detection of the pair structure of leptons by way of the discovery of the muon neutrino."

The Lederman, Schwartz, Steinberger method for creating neutrino beams was among the first to use a particle accelerator to generate laboratory-made neutrino beams. Their method opened a new window into measuring neutrino decays and neutrino interactions with matter. In 2023, neutrino physics remains a very active area of research.

Antiprotons and Antineutrons

By the 1950s, physicists were theorizing that each elementary particle had a corresponding antiparticle of opposite charge but identical in all other respects. This concept became known as charge symmetry. In 1928, Dirac had shown mathematically that when Einstein's STR was included in his field equations of quantum mechanics, every particle had a corresponding antiparticle. The antiparticle would have the same mass but opposite charge. The discovery of the positron in 1932, which had the same characteristics as an electron but exhibited a positive charge, was thought to be an example of this symmetry. Later, with the discovery of positive and negative pions and muons in cloud chambers, photographic emulsion stacks, and cyclotron experiments, this hypothesis was reinforced.

In 1954, Ernest Lawrence organized the construction of the weak-focusing proton synchrotron named the Bevatron at the Radiation Laboratory at the University of California Berkeley. Earlier, in 1939, Lawrence had been awarded the Nobel Prize in Physics "…for the invention and development of the cyclotron and for results obtained with it, especially with regard to artificial radioactive elements." He was an expert in building particle accelerators. Now, the motivation for the Bevatron was to confirm the hypothesis of charge symmetry and prove, in this case, that the antiparticle of the proton existed. The Bevatron's beam energy was designed to be 6.2 GeV; enough energy to create a proton-antiproton pair when protons were made incident on a fixed target.

In 1955, the antiproton was soon discovered by a team of physicists at the Bevatron led by the Italian-American Emilio Segrè (1905–1989) and the American Owen Chamberlain (1920–2006). The experiment identified the antiprotons from the scattered particles from the target, directing them through a shielding wall surrounding the accelerator into a 15 m channel made of dipole and quadruple magnets. At both ends of the channel were scintillation counters, which measured their time-of-flight between the ends. The antiprotons were separated from the large background of accompanying negative pions by the measurement of the particle's flight time and momentum. By years end, antiprotons had been identified. A year later, the antineutron was found by another team led by the American physicist Bruce Cork (1916–1994). In 1959, the Nobel Prize in Physics was awarded to both Segrè and Chamberlain "…for their discovery of the antiproton."

Charge Parity Violation

For many years, physicists thought that at the level of elementary particle interactions, under the transformation of a spatial inversion (parity transformation P), or the substitution of antiparticles for particles (charge reversal C), or the reversal of time (T) the results would be the same. In 1956 the weak interactions, however, were found not to be invariant under the spatial inversion P. For a few more years, physicists were inclined to accept this asymmetry. Although parity symmetry was broken, the violation of P symmetry could be compensated for by a simultaneous reversal of charge symmetry C in the reactions. If one were to simultaneously consider both charge and parity transformations, CP would be the transformation that was conserved. This was was correct until this CP symmetry was also found to be violated in neutral kaon decays, although at a magnitude much smaller than the violation of parity was.

The neutral kaon was discovered in 1947 in cloud chamber experiments where it decayed into two charged pions. In 1956, Lederman's team on the newly commissioned Brookhaven AGS, discovered that the neutral kaon decay occurred in either of two quantum states. In one, referred to as the K-short, it decayed weakly into two pions with a short lifetime of 0.9×10^{-10} s; in the other, referred to as the K-long, it decayed into three pions with a longer lifetime of 0.5×10^{-7} s.

The neutral kaon, with a mass of 493.3 MeV/c^2, was only slightly above the mass of its three-pion decay mode. This slight mass difference contributed to its longer lifetime relative to the two-pion decay mode. The K-short, with its symmetric two body pion decay, existed in a weak quantum state having a value of CP $= +1$, whereas the K-long, with its three-body decay, had the opposite value of CP $= -1$.

In 1964, while trying to verify the results of another experimental group, an American experimental team headed by James Cronin (1931–2016), Val Fitch (1923–2015) with the student James Christenson (1937–) and the French nuclear physicist Rene Turlay (1932–2002) fielded a precision experiment to measure the neutral K-long decays produced at the AGS. The neutral K beam was generated by impacting an internal Be target with the 30 GeV protons from the accelerator. The neutral kaons subsequently passed freely into a beam channel of 17.4 m length, becoming a pure K-long beam at its exit; while the K-shorts had all decayed away during propagation.

At the channel's end, decays of the remaining kaons were measured in a disymmetric magnetic spectrometer system composed of scintillator counters and spark chambers. During the experiment, decays of 22,700 K-long decays were measured. Unexpectedly, within the decays, 45 two-pion K-short events were observed. These decays violated CP symmetry, one part in 500.

An intensive study of CP violation symmetry consumed many physicists, theorists and experimentalists in the following decade. During this period, Steinberger's team at the AGS and later at CERN established the definitive measurements of the neutral kaon's CP violating decay parameters, while trying to bring understanding to this violation.

That CP was a broken symmetry was contrary to establish theory; no appropriate alternative was available to replace CP invariance, as in parity violation with the simultaneous violation of charge reversal. There was also the question of the CPT theorem central to the validity of quantum fields theories, which stated that nature was symmetric under the simultaneous transformation of all three transformations: C, P and T. If CPT holds, then CP violation implies violation of the revered time reversal symmetry T. The discovery of CP violation thus led to the conclusion that the microscopic laws of physics allowed absolute distinctions between right-handed and left-handed co-ordinate systems, between antiparticles and particles, and between time running backwards or forwards.

In 1980, the Nobel Prize for Physics was awarded to James Cronin and Val Fitch "…for the discovery of violations of fundamental symmetry principles in the decay of neutral K-mesons." The experiment implied that reversing the direction of time would not precisely reverse the course of certain reactions of subatomic particles. According to Big Bang theories, at the time of the Big Bang antiparticles and particles were assumed generated in equal numbers. The present universe, contains nearly no antimatter. The asymmetry in CP violation theorizes a possible mechanism.

Hadrons and Quarks

The Bubble Chamber

The detection and measurement of particles using cloud chambers and photographic emulsion stacks in the newly constructed particle accelerators quickly became inadequate as both were time consuming. Especially tedious was the use of emulsion stacks, which were scanned by a microscope. Cloud chambers needed time to reset between particle interactions and both could not keep pace with the accelerators rate of particle production.

In 1952, the bubble chamber, invented by the American physicist Donald Glaser (1926–2013) resulted in a partial improvement over these difficulties. It was comprised of a chamber filled with a liquid heated to just below the boiling point. As particles entered the chamber, a piston suddenly decreased the chamber pressure and the liquid entered a superheated state. Charged particles generated an ionized track around which the liquid vaporized, resulting in bubbles. Multiple cameras that surrounded the chamber permitted a three-dimensional image of these interactions to be photographed. If the chamber was also surrounded by a magnetic field, a measured radius of curvature of a charged particle track enabled its momentum to be determined.

Within several years of its invention, the bubble chamber replaced the use of the emulsion stack and cloud chamber as a particle detector. Its development lead the way for many discoveries in the arena of particle physics. In 1960, Glaser was awarded the Nobel Prize in Physics "…for the invention of the bubble chamber."

New Resonance States

In 1953, the bubble chamber was quickly advanced by the American physicist Luis Alvarez (1911–1988) after becoming acquainted with Glaser's work. Alvarez was convinced that the chamber filled with liquid hydrogen would be a much improved method for tracking particles exiting accelerators. By 1954, he had assembled a small liquid-hydrogen bubble chamber and by 1956 a larger one was built at the Bevatron. The chamber was cycled in synchronization with the accelerator beam, with photographs taken and the chamber recompressed in time for the following cycle of beam particles.

Alvarez's efforts concentrated on development of high-speed machines to measure and analyze the millions of photographs produced by an assembled complex of machines. Thousands of particle interactions were measured and analyzed in detail, resulting in the discovery of a large number of short-lived particle resonances. Hundreds of the new particles and their excited states were identified. In 1968, Alvarez was awarded the Nobel Prize in Physics "...for his decisive contributions to elementary particle physics, in particular, the discovery of a large number of resonance states, made possible through his development of the technique of using hydrogen bubble chambers and data analysis."

Luis Alvarez is also known for the Alvarez Hypothesis, which suggested that the extinction event of 66 million years ago that destroyed the non-avian dinosaurs, was the result of an asteroid impact. This hypothesis was developed together with his geologist son Walter Alvarez in 1980 and presented at a formal CERN lecture, which the author was fortunate to attend. His eleven arguments for the hypothesis were cogent and convincing, despite the fact that the impact location had yet to be discovered. In the early 1990s, the location was identified as a 300 km crater located adjacent the Yucatan Peninsula in the Gulf of Mexico. That discovery, affirming the impact date of 66 MYA, gave immediate credibility to their hypothesis.

The Quark Model

The new particles seen in bubble chambers were hadrons; particles sensitive to the strong force that holds the nucleus together. These consisted of nucleon and meson-like particles. Most hadrons had short lifetimes varying from 10^{-23} to 10^{-8} s. The numerous particles discovered suggested that there must be an underlying structure, likely composed of smaller and fewer elementary particles.

In the early 1960s, two theoretical physicists, American Murray Gell-Mann (1929–2019) and Israeli Yuval Ne'eman (1925–2006), independently observed that the Lie algebra group of dimension three, SU(3) provided a mathematical model that correlated with many of the quantum numbers of the newly discovered hadrons. When plotted, using their quantum numbers, the hadrons formed simple geometric patterns. Because SU(3) is a compact eight parameter Lie group, the Gell-Mann/

Ne'eman model was named the "Eightfold Way." This formulation was not based on any fundamental theory. Its practicality was exhibited in its short representation expressing order and its symmetry with predictive power.

The basic structure of the hadrons was described as composites of just a few basic particles, referred to as quarks. Quarks were posited to be of three types and referred to as flavors: up, down and strange. All quarks were assigned a baryon number of 1/3 and a spin of 1/2 and referred to as fermions, named after Enrico Fermi. Up quarks had an electric charge of +2/3, while down and strange quarks had an electric charge of −1/3. Antiquarks had the opposite quantum numbers. Baryons were the name given to the nucleons and new nucleon-like particles. Baryons were made of three quarks and accordingly have baryon number one. Mesons were made of quark-antiquark pairs and accordingly have baryon number zero.

The quarks were therefore the true elementary particles and the fundamental constituents of matter. They merged to form the composite particles called hadrons. Mass differences among the hadrons were due to different masses of the constituent quarks. Protons and neutrons were the most stable hadrons. The proton was viewed as composed of two up quarks and one down quark; the neutron was one up quark and two down quarks.

Initially, there were several problems with the Quark Model. First, physicists had never detected a quark. Second, to make the theory of quarks function properly, they had to be made up of a fractional electric charge. All particles previously discovered had only charges of whole numbers: 1, 0, or −1. In 1964, some credibility was given to this Quark Model. A major prediction claimed the existence of a baryon named the omega-minus. It was characterized as a particle with the following quantum numbers: baryon number + 1, spin 3/2, positive parity, negative charge, strangeness −3 and a mass of 1680 MeV/c^2. The same year, a team of physicists from the University of Rochester, the University of Syracuse and Brookhaven National Laboratory led by Nicholas Samios (1932–), using the 80 in bubble chamber at the AGS, discovered the predicted omega-minus, thereby completing the mathematical structure. In 1969, Gell-Mann was awarded the Nobel Prize in Physics "...for his contribution and discoveries concerning the classification of elementary particles and their interactions."

By the early 1970s, additional evidence for the quarks emerged in deep inelastic-scattering experiments at SLAC (Stanford Linear Accelerator Center). These experiments showed that the proton and neutron contained much smaller, point-like objects, and therefore were not elementary particles. Conceptually, these experiments were similar to those of Rutherford's years ago, where the nucleus was discovered inside the atom. In contrast, in these experiments the projectiles were electrons having energies up to 20 GeV generated by the Stanford Linear Accelerator, a 3.2 km long accelerator, constructed in 1966, and the target was liquid hydrogen. Despite the greater spacial resolution, physicists were not completely convinced that these objects were quarks. Their charge was not determined and their size was too small to be measured. They were simply labeled partons by physicist Richard Feynman.

In 1990, three key members of the SLAC-MIT team who discovered the partons, the Americans Jerome Friedman (1930–), Henry Kendall (1926–1999) and Richard

Taylor (1929–2018) were awarded the Nobel Prize in Physics. The citation read "…for their pioneering investigations concerning deep inelastic scattering of electrons on protons and bound neutrons, which have been of essential importance for the development of the Quark Model in particle physics."

More Quarks and Leptons

Charm Quark

In 1967, physicists at the AGS developed the ability to efficiently extract its 30-GeV proton beam and direct it to an external target, thus providing a significant gain in secondary particle production. With this new tool, motivated by Lederman, his Columbia University team developed a very sensitive way to look for intermediate vector bosons as well as other new particles when beam protons impact a target. The vector boson was a heavy particle, predicted to facilitate the neutron transition to protons in radioactive beta decay. Lederman's key was to search for possible rare, high transverse-momentum muon pairs ($\mu^- \mu^+$). Such pairs might occur as a bump in a monotonic decreasing yield spectrum of muon pairs with mass. This type of enhanced pairing was predicted to be a distinguishing feature of vector boson decays.

In Lederman's experiment, the target used was uranium. The muon momentum was determined by measuring their penetration range in ten feet of steel shielding interspersed with liquid and plastic scintillation counters. This instrumentation permitted high data rates to be achieved for the first time. Because of the enormous flux of 10^{11} protons/pulse, the experiment was very sensitive to small yields. Signals were recorded to a level of 10^{-12} of the total cross section. Over the mass range explored from 1 to 6 GeV/c^2, the team observed an enhancement near 3.5 GeV/c^2. It formed as a broad shoulder on top of a seven order-of-magnitude, exponentially decreasing spectrum of muon pairs with mass. With limited energy resolution, resulting from the MCS (multiple coulomb scattering) of the muons in the high Z target and steel, little could be inferred except that the enhancement observed might possibly be due to a composite resonance.

Although little new was learned, the experimental technique that had been developed was a crucial feature for this class of super high-rate data collection experiments. It provided a reliable way of reducing the background when searching for rare events.

In 1974, an MIT group directed by the American physicist Samuel Ting (1936-) repeated this 1968–1969 experiment at the AGS, this time searching for rare events having electron positron pairs ($e^- + e^+$) produced from a beryllium target. The detector employed a magnetic spectrometer based on a 8000-wire proportional chamber technology. The structure of the enhancement was refined by reduced MCS and superior resolution of the chambers into a soaring peak at 3.1 GeV/c^2, with a narrow width of 0.1 GeV/c^2, called the J particle. The width of the J particle's mass peak was 1000 times narrower, with an associated lifetime 1000 times longer than other particles of comparable mass. There was significant theoretical speculation as to what this unique, unexpected discovery was. To assist in providing an explanation of the peak, the experiment expanded to include related searches to look for more J particles, in which the author was a participant.

At SLAC on the West coast, a different type of experiment was proceeding. There, a SLAC-Berkeley group, under the leadership of the American physicist Burton Richter (1931–2018), was conducting experiments on Stanford's SPEAR (Positron–Electron-Asymmetric-Ring) collider. SPEAR consisted of a single ring, 80 m in diameter, in which counter rotating beams of electrons and positrons revolved at energies up to 4 GeV. Particle collisions were measured in a 4 kg magnetic solenoid detector, surrounding a region where the two beams collided. The detector contained a system of wire spark chambers and scintillation counters.

SPEAR was built to study electron positron collisions ($e^+ \times e^-$) in a new energy region above 3 GeV, higher than had previously been studied with lower energy e^+e^- colliders. The initial plan was to collect data at a number of different collision energies spanning the new energy region. In the process of doing these energy scans, a large peak in the energy spectrum was found at an energy just below 3.1 GeV, indicating the production of a new particle, with a narrow width and large production rate. That particle was referred to as the ψ (psi), by the SLAC-Berkeley group. The production mechanism was the inverse of that used to produce the J particle at Brookhaven.

This unexpected discovery of the same particle as the J particle, at two different laboratories and using completely different experimental methods, at the same time, was noteworthy. Especially notable were the unique properties of this particle that were very different from properties observed in particles previously. Because the discoveries of the enhancement were considered to be done independently, the particle community used both names. It became referred to as the J/ψ particle.

The question remained as to what the new particle was. It elicited sensational excitement once it was announced jointly by both groups on November 10, 1974: the MIT group conducting the experiment on the East Coast at the Brookhaven National Laboratory and the SLAC-Berkeley group performing the inverse experiment on the West Coast at the Stanford Linear Accelerator Center. Its unprecedented characteristics, on two entirely different machines, using two different techniques, led to a host of speculations and vigorous experimental activity. Over the following year, greater than 700 scientific papers relating to the observation were written; a record in physics. In the world of high energy physics, that time was known as the November 1974 Revolution.

Subsequent exploration by the SPEAR team exhibited the existence other ψ states, the ψ' at 3.684 GeV/c^2 and the ψ'' at 3.770 GeV/c^2. With time, clarity prevailed. The J/ψ was a bound state of a new quark called charm and its antiparticle called charmonium. The states above the J/ψ mass were the excited states of a charmonium meson. The mass structure of charmonium was similar to positronium, a bound state of an electron–positron pair where the two particles rotate around one another similar to an atom. Positronium was predicted around the time the positron was first discovered in 1932. In 1951, it was finally confirmed by the Austrian-American physicist

Martin Deutsch (1917–2002). The charmonium energy spectrum was similar to that of positronium's excited states. The J/ψ was now understood, within the structure of non-relativistic quantum mechanics, of two particles (one charmed and the other anti-charm) revolving around each other. In 1976, the Nobel Prize in Physics was awarded jointly to Samuel Ting and Burton Richter "...for their pioneering work in the discovery of a heavy elementary particle of a new kind."

Seven years later, charmed D-mesons composed of the new charmed quark together with combinations of up, down or strange quarks were looked for and discovered. In one of the first successful searches at CERN, the author again participated. The accidental discovery of charmonium in its different manifestations, at last, gave the Quark Model an experimental credibility.

The use of MWPCs (multi-wire proportional chambers), such as those used by the Ting group and by other experimenters at that time, demonstrated the utility in the quantum-jump in recording the production rates. The new MWPC technology was coupled with the expanding developments in computers. This technology allowed for the discovery and study of rare events, like those containing charm, which needed high statistics and precision resolution. The bubble chamber and spark chamber detectors could only record events at rates less than a few per second.

In contrast, the MWPC and associated drift chamber technology allowed particle interactions to be recorded at microsecond rates. These high rates enabled new particle discoveries. By 1979 they became the dominate detectors employed in many of the cutting-edge particle physics experiments. In 1992, the Polish-French physicist Georges Charpak (1924–2010) was awarded the Nobel Prize in Physics "...for his invention and development of particle detectors, in particular the multi-wire proportional chamber."

Tau Lepton

Compared to the accidental discovery of the charm quark, the search for a lepton, heavier than the electron and muon, had been underway for some time. The first experiments to find the lepton were done in 1973 at the ADONE electron–positron collider at Frascati, Italy. ADONE had a ring diameter of 33 m and a beam energy of 1.5 GeV, with a center-of-mass energy of 3 GeV when its electron–positron collided.

In the following year, the American physicist Martin Perl (1927–2014) and his colleagues, continued the lepton search using the larger diameter and higher energy SPEAR storage ring. In 1975, immediately following the J/ψ discovery, Perl announced that they had found a lepton, which they named the tau τ. In the experiments, they did not detect the tau directly, but indirectly through events, where the tau was inferred through the production and subsequent decay of a new particle pair $(\tau^+ + \tau^-)$. The decays were measured by the path: $e^+ \times e^- \rightarrow \tau^+ + \tau^- \rightarrow e^\pm + \mu^\mp + 4\nu$. The ν represented the neutrinos that were undetected. The tau mass was estimated to be 1.79 GeV/c^2, approximately 17 times the mass of a muon, which was more than 200 times that of an electron. In 1996, the Nobel Prize was awarded jointly to Perl "...for pioneering experimental contributions to lepton physics" and to Reines "...for the detection of the neutrino."

The consensus, within the particle physics community, was that for every lepton there would likely be an associated neutrino, as this was the case for both the electron and muon. Twenty-two years later, in 1997 the tau neutrino was found and measured at Fermilab, completing the sequence of three lepton pairs: electron, muon and tau with their associated unique neutrinos.

Bottom Quark

Concurrent with the discoveries of the charm quark and the tau lepton, experiments were underway at Fermilab where the next two quarks were discovered. In the early 1960s, the laboratory, which first began as the National Accelerator Laboratory (NAL), had been promoted by Lederman. Since his award-winning discoveries, Lederman became a tireless supporter of science. As he gained stature, he was able to influence science policy, including his advocacy of NAL.

NAL construction broke ground west of Chicago, Illinois, in 1968 under its first director and architect, the American physicist Robert Wilson (1914–2000). By 1972, NAL had accelerated a proton beam in its 6.3 km circumference main ring to the design energy of 200 GeV. By the end of 1973, the accelerator system was operating routinely at 300 GeV, and by 1976, it had reached energies of 500 GeV. In 1974, NAL was renamed the Fermi National Accelerator Laboratory and referred to as Fermilab. The name honors Enrico Fermi whose achievements, placed him among the great physicists of the twentieth century.

In 1974, Lederman continued his search for new particles having larger masses, using the higher beam energy available at Fermilab and expanding on techniques he pioneered at the AGS in 1969. With the new 400 GeV proton beam at Fermilab, his group developed an experiment, again using oppositely charged muon pairs, in two massive muon detectors located downstream that were placed to the left and the right of the target beam line. Upstream of the detectors was a tungsten beam dump, surrounded on either side by two 9 m long beryllium absorbers to remove unwanted hadrons incident on the detectors. A steel shielding wall with two apertures allowed the muons to pass into the detector's arms. Each arm consisted of a large analyzing magnet, followed by three planes of MWPCs, a thick magnetized steel block, three MWPCs, a thinner steel absorber and two more planes of MWPCs.

The decreased MCS and associated increased resolution of the instrumentation improved particle identification and mass resolution of the muons from 10–15% of the 1968 Brookhaven National Laboratory experiment to 2%. The filtering of the hadrons enabled over a thousand times as many protons to hit the target as earlier, thereby greatly improving the event statistics. The result was a clean exponential spectrum of muon pairs accumulated over the mass range 5 to 14 GeV/c^2, where the measured cross section was scaled from 10^{-15} to as low as 10^{-39} cm^2/GeV. Over the rapidly diminishing cross section with increasing mass, a clean enhancement of 800 events emerged at a mass of 9.5 GeV/c^2. The resonance was clear and named the upsilon Y (9.46 GeV/c^2). The discovery was announced on June 30, 1977 and by September, with 30,000 events, the enhancement resolved into three well-established peaks. The resonances appeared more and more like those of the J/ψ system but at much higher mass.

The DORIS electron positron collider at DESY (Deutsches Elektronen-Synchrotron) in Hamburg, Germany, built in 1974, produced the upsilon from electron positron collisions, indicating that the only interpretation of the resonance, was that of a bound state of a new quark with its antiparticle. The new quark was called bottom and labeled **b**. The upsilon was a new heavy meson made of the bottom quark and its antiparticle revolving around each other, similar to the J/ψ meson with its charm quark and antiparticle revolving around itself. As in the study of charmonium, the upsilon meson provided an additional laboratory for discovering the details of the strong force that holds the quarks together.

The Lederman team of 16 members came from three institutions: Columbia University, Fermilab and the State University of New York at Stony Brook. That team represented one of the last small groups that made major discoveries in particle physics. The complexity of the experiments, as exemplified by those being conducted on colliding-beam accelerators, requires larger teams of physicists working together from more institutions. This increase in complexity is best illustrated in the search for the top quark at Fermilab.

Top Quark

By 1978, two families of particle matter had been identified. The first consisted of two up and down quarks and two leptons, the electron and the electron neutrino. The second consisted of the charm and strange quarks and two leptons, the muon and the muon neutrino. With the discovery of the bottom quark and tau lepton and the tau neutrino, a third family was postulated. In that third family, a possible new quark, referred to as the top quark, was expected to balance the bottom quark, as the tau neutrino balanced the tau lepton. Accordingly, there was new motivation to search for this quark.

In 1979, Lederman became director of Fermilab. That year he marshaled the support of the particle physics community and the science advisor to then President Reagan to the concept of building an energetic proton antiproton collider at Fermilab to assist in putting a foundation on this formulating model of matter. In that year the collider was also approved and finding the top quark was high on its agenda.

In due course, the accelerator complex became known as the Tevatron. It was completed in 1983 under the leadership of Lederman and the American physicist Alvin Tollestrup (1924–2020) at an initial cost of $120 million. By late 1986, two beams of 900 GeV particles, one with protons and the other with antiprotons were circulating at the super-conducting Tevatron, producing collisions of 1.8 TeV. The enormous energy produced at the collision point introduced a new scale in energy generation, the TeV (Tera electron volt) equal to 1000 GeV.

Concurrent with the Tevatron development, two large particle detectors were built. The first one built was the CDF (Collider Detector at Fermilab), located at one of the collision regions of the Tevatron. Later, the DØ (DZero) detector was constructed at the opposite side of the ring with differing detector arrangements and analysis techniques. These two detectors allowed independent confirmation of each others discoveries. The detectors were massive in size and filled with complex instrumentation. CDF spread over 12 m in all three directions, from the beam's intersection region. It was composed of six layers of intricate detectors in a structure weighing 5000 tons. DØ was equally large and complex.

The design and construction of the particle detectors were finalized by two separate international collaborations, requiring hundreds of physicists who analyzed the billions of events recorded. The CDF collaboration was composed of 600 physicists from 30 American universities and national laboratories and an equal number from foreign universities and laboratories. The DØ collaboration was similarly formed; it was composed of 650 physicists from 88 universities and national laboratories from 21 countries.

In February 1995, the extensive search for the top quark concluded with its simultaneous publication in the *Physical Review Letters* by both experimental teams. The data showed that the top quark t had the very high mass of 175 GeV/c^2 in both the CDF and DØ detectors. Although the existence of the top quark was expected, its enormous mass was a surprise. After the earlier discoveries of the charm and bottom quarks with masses of 1.5 GeV/c^2 and 4.5 GeV/c^2, respectively, simple numerology raised expectations that the top mass would be about 13.5 GeV/c^2. Finding the top mass at more than 10 times this value was unforeseen. In addition to the top quark production, via the top anti-top quark process, top quarks were also produced via the top anti-bottom quark process, thereby allowing additional measurements of the top quark properties. With these measurements at the Tevatron, the third family of particle matter was confirmed.

Fourteen years later, both the CDF and DØ collaborations announced their discovery of the generation of single top quarks by the weak interaction. This mechanism occurred at approximately half the rate as for the production of the top quark pairs. A single top quark was more difficult to find because of trouble separating it from background processes. These studies were used to measure the top quark lifetime of 5×10^{-25} s, as well as precision measurements of its charge, mass, decay modes and production characteristics. All of which were reported in more than 100 publications. In 2019, the European Physical Society awarded its High Energy and Particle Physics Prize to the CDF and DØ collaborations "...for the discovery of the top quark and the detailed measurement of its properties."

Gluons

The gluon was predicted as a spin-one elementary particle that acts as an exchange particle between quarks, binding them together by the strong force. It ties quarks together, forming hadrons such as protons and neutrons in the nucleus. Its function is analogous to the exchange of photons between two charged particles when they interact by the electromagnetic force. In contrast to the force between charged particles and gravitational masses that decrease with increased particle separation, the strong force increases with separation. This is similar to the concept of a rubber band; the harder pulled apart the more it pulls back and it also breaks. As a result, single quarks are never observed. They always appear together with other quarks or antiquarks in combinations of pairs, as in mesons or as triplets in baryons. When quarks emerge from a collision with high enough energy, they can pull new particles out of the vacuum to become a cloud of particles. (Energy is distance times force and that energy can be converted into mass.) An isolated quark, therefore, doesn't stay isolated; it produces less energy to create more quarks than to be single.

For a highly energetic quark or gluon flying away from a collision, this process happens several times, resulting in a narrow jet of hadrons that have the same direction as the original quark. For lower energy quarks, the resulting quark byproducts of the collision spread out, becoming indistinguishable from the original projectile's direction. Thus, it was not until very energetic particle accelerators became available that individual quark jets were finally observed.

In 1976, a number of theoretical physicists suggested looking for a gluon in three-jet events that are due to gluons forming in electron positron collisions. In the collision, two quarks (and their associated hadronic jets) form, following the original electron positron directions. The radiated gluon, separated from the two quarks, would form a third hadronic jet.

In 1979, using this method, the gluon was discovered by the PLUTO collaboration on the PETRA (Positron-Elektron Tandem Ring Anlage) collider at DESY, which had energies ranging from 10 to 45 GeV. The collaboration was comprised of 35 German physicists from five German Institutions together with physicists from universities in Great Britain, Italy, Israel and the United States. Measurements confirmed its spin-one and massless characteristics. In 1995, the Special High Energy and Particle Physics Prize of the European Physical Society was awarded to the PLUTO collaboration "…for establishing the existence of the gluon in independent and simultaneous ways."

Intermediate Vector Bosons

Prediction

In 1687, Newton mathematically unified the seemingly disparate forces of falling objects on the Earth with those controlling the planets, using the Law of Universal Gravitation. In 1873, Maxwell mathematically unified the seemingly different forces of electricity and magnetism, showing they were both manifestations of the same electromagnetic force. Now in 1967, the American physicist Steven Weinberg (1933–2021) theoretically unified the electromagnetic force with the weak force of nuclear beta decay into a mathematical electroweak theory.

Electromagnetism operates at potentially infinite distances by means of the exchange of the massless particles called photons. The weak force operates within the domain of the nucleus. Quantum theory suggests that the range of a force varies inversely with the mass of its messenger particle. Thus, in contrast to the massless photon, Weinberg predicted that the weak force in the nucleus would occur by the exchange of massive charged particles referred to as intermediate vector bosons: the W^+, W^- and the neutral Z°. The predicted weak force would produce the radioactivity of beta decay of the neutron and the resulting transmutation of the nucleus. The weak force would also play a key role in the solar furnace within the Sun, which is powered by the nuclear fusion of hydrogen into helium. All transformations would come about as the result of the actions of both the strong and a new electroweak force.

Weinberg predicted that despite their apparent dissimilarities, the massless photon and the three massive bosons were in effect members of the same mathematical group of four particles. During the early moments of the Big Bang, within the very high energies produced, electromagnetism and the weak forces were assumed to be indistinguishable. This theory was modeled on concepts developed for QED.

During the mid-1960s, a similar mathematical unification was being developed independently by the theoretical physicists American Sheldon Glashow (1932-) and the Pakistani Abdus Salam (1926–1996). Similar to Weinberg, they made predictions of new particles, like the Z^0, a neutral intermediate vector boson that could act as an exchange particle between a muon neutrino and a stationary electron producing an outgoing muon neutrino and an energetic electron. This reaction was called a neutral current because the Z^0 carried no electric charge. They developed a new class of unified quantum field theories that described the two forces of nature (electromagnetism and weak), based on the exchange of particles of the fields, which appeared prior to that time to be independent of each other. Important in their construction was that the theory would be built with an underlying mathematical symmetry where the effects of the force are the same at all points in time and space. Prior to this time the theories and conjectures were only mathematical.

Bosons described particles that have spin one, like the photon or the gluon. As spin one force carriers, they were referred to as vector bosons. The boson was named after the Indian theoretical physicist Satyendra Nath Bose (1894–1974). The name parallels that of fermions, named after Enrico Fermi, which have spin one-half and characterize quarks and leptons. The vector bosons had previously been searched for, by the Ting and Lederman teams in experiments where they discovered, accidentally, instead, the charm quark and the anticipated bottom quark, respectively, but no vector bosons.

While the discovery of the new quarks was being pursued in the USA at Fermilab, a major effort was also underway at Europe's CERN to investigate the properties of this weak interaction. In the process, a friendly competition developed between the two laboratories to see who could find the next important physics discovery first.

Indication at Gargamelle

In 1970, a very large bubble chamber was built at CERN. The chamber was designed to detect neutrinos produced in a muon neutrino beam generated by 26 GeV protons from the CERN PS (Proton Synchrotron). The bubble chamber, named Gargamelle, was 4.8 m in length and 2 m in diameter. It weighed 1000 tons and held 12 cubic meters of heavy liquid freon. It was surrounded by a magnetic field of 19 kg, with its yoke weighing 800 tons and the magnet enabling the momentum of any neutrino created charged particles to be measured with precision. In July 1973, the Gargamelle collaboration presented in a CERN seminar the first indirect evidence of the predicted weak neutral current carried by a Z^0 boson. The mechanism was as described in the above hypothetical example by Glashow and Salam. In September 1973, the collaboration published two associated papers in *Physics Letters,* the premier European physics journal. This discovery gave CERN its first major particle discovery. In 1979, the Nobel Prize in Physics was jointly awarded to Glashow, Salam and Weinberg "…for their contributions to the theory of the unified weak and electromagnetic interaction between elementary particles, including the prediction of the weak neutral current."

Discovered at SPS Collider

In 1976, the SPS (Super Proton Synchrotron), became the succeeding accelerator to the PS at CERN. The PS was now used as an injector to the SPS, which increased the beam energy of its 26 GeV protons to 300 GeV in the 6.9 km circumference beam tunnel of the SPS. This energetic proton beam at CERN expanded to encompass the border of Switzerland with France adjacent to Geneva.

The same year, the Italian physicist Carlo Rubbia (1934–), together with his colleagues, proposed searching for the remaining particles predicted by the electroweak theory, namely the charged W^+ and W^- intermediate vector bosons and also the Z^0, in a creative new way. All three particles had been postulated to transmit the weak force in the nucleus. The Ws were responsible for the charged-currents in the weak interaction. There, an up quark could combine with an anti down quark to form the W^+ (positive current), to produce a positron and an electron neutrino. These particles, like the Z^0, were estimated to weigh in the range of 100 GeV/c². They were expected to be very massive as their range was constrained by existing within a nucleus.

When protons collide with a fixed target, the energy generated in the particles' center-of-mass system is degraded by the square root of the beam energy and therefore is not be able to reach the required energy at the SPS. By colliding the protons directly with antiprotons, however, these energies are reachable. Rubbia thus suggested upgrading the SPS from a one beam accelerator to a particle collider with two beams. Colliding protons with antiprotons could create enough energy to produce both the Ws and the Z^0 bosons. The breakthrough that made the idea feasible was the invention by the Dutch accelerator physicist Simon van de Meer (1925–2011) of cooling methods for the antiprotons. His discovery permitted the beam of antiprotons emerging from a target to be compressed in both transverse dimensions and momentum, making collisions with on coming protons possible.

Despite the technical challenges, during 1980–1981, the SPS was modified to a proton-antiproton collider for a limited time. With a chosen beam energy of 270 GeV per beam, an energy of 540 GeV in the center-of-mass was sufficient to produce the bosons. Simultaneously, two very large particle detectors UA1 and UA2 were built at opposite ends of the collider to look for and measure the decay characteristics of these bosons. These detectors were developed by two independent groups of collaborators. Rubbia was the spokesperson for UA1, with the French experimental particle physicist Pierre Darriulat (1938–) becoming the spokesperson for UA2. The rationale for two colliders and the nature and size of the detectors and collaborations were similar in motivation and scope to the two detectors to be built at Fermilab looking for the top quark.

The first tests of the collider began in June 1981. From October through December 1982, the SPS operated as a collider for boson physics experiments. By June 1983, the decays of tens to hundreds of $W^\pm \to \mu^\pm \pm v_\mu$, $W^\pm \to e^\pm \pm v_e$, and $Z^0 \to e^+ e^-$ were measured, within the billions of particles generated. These events allowed the masses of the W and Z^0 bosons to be established at 80 GeV/c² and 90 GeV/c², respectively. Both masses were in agreement with the electroweak model. The three particles were short lived, having a half-life of 3×10^{-25} s.

Using the electroweak theory, the radioactive beta decay, discovered by the Curies and Rutherford at the turn of the nineteenth century, was finally understood as a two-step process involving the weak decay. At one level, the neutral neutron within the nucleus transforms into a positive proton and a negative electron, plus an anti-electron neutrino: $n^0 \to p^+ + e^- + v_e$. But at the quark level, the decay is a two body sequence. First, within the neutron (which is made of a conglomerate of three quarks: two down-quarks and one up-quark) one of its down-quarks $d^{-1/3}$ decays into a negative W^{-1} and an up-quark $u^{+2/3}$: $d^{-1/3} \to W^{-1} + u^{+2/3}$. The remaining down and up quark plus a transformed up quark, within the original neutron, form the proton (made of a composite of the one down-quark and two up-quarks). Secondly, the W^{-1} decays into an electron and the anti-electron neutrino: $W^{-1} \to e^- + v_e$. Throughout these transformations, the electric charge is conserved.

In 1984, Rubbia and van der Meer shared the Nobel Prize in Physics "…for their decisive contributions to the large project, which led to the discovery of the field particles W and Z, communicators of the weak interaction."

Detailed Measurements at LEP

By 1980, and prior to the above measurements of the vector bosons, a new structure of the fundamental interactions unifying the electromagnetic and weak forces was emerging, leading to an embryonic SM (Standard Model of Particle Physics). In 1981, physicists from CERN's member states made the decision that a Large Electron Positron Collider (LEP) would be CERN's next flagship machine to study electroweak interactions with precision. In contrast to CERN's past focus on proton accelerators, the decision was made to accelerate electrons and positrons. The results of such collisions would be easier to interpret than those between protons and antiprotons. The leptons (electron and positron) were single particles with no internal structure to confuse any interaction. The hadrons (proton and antiproton), in contrast, were made of quarks complicating the interpretation of any collision. A beam energy of 45 GeV was chosen so that the Z^0 boson, with its estimated mass of 90 GeV/c^2, could be achieved.

In 1981, the LEP was formally approved. Construction began in September 1983 when the presidents of CERN's two membership countries, Pierre Aubert (1927–2016) of Switzerland and François Mitterrand (1916–1996) of France, symbolically broke soil and laid a plaque commemorating the inauguration.

The biggest component of LEP was its 27 km circular tunnel and its ring of magnets where the beams achieved full energy. The tunnel and experimental detectors were located 100 m below ground on a plane with a 1 degree tilt to avoid the underground water springs that existed under the Jura Mountains, between Switzerland and France. The magnetic ring was injected with electrons and positrons traveling in opposite directions, delivered by CERN's accelerator complex. These particles were first generated and accelerated by a pre-injector. They, then were accelerated to almost the speed of light by the PS, followed by the SPS where they were injected into the LEP magnetic ring. With its 27 km circumference, the LEP collider was the largest electron–positron accelerator ever built. Its excavation was Europe's largest civil engineering project prior to the boring of the Channel Tunnel.

In February 1988, the two sections of the magnetic ring came together with a misalignment of just one centimeter. The first beam successfully circulated in July 1989, with collisions being measured a month later. On November 1989, LEP was officially inaugurated. Present at the inauguration were ministers and heads of state from all of CERN's 14 member states, together with 1500 guests.

During LEP's construction, four immense detectors Aleph, Delphi, Opal and L3 were assembled in underground halls around four collision points of the collider. The size of each was similar to a small multi-story house. As with the recently completed Tevatron detectors at Fermilab, these were designed independently with differences that allowed for complementary experiments. The components of the detectors came not only from CERN's member states but also from the USA, China and Japan. This particular group of physicists became the most international collection of scientists the world had ever seen, working together in collaboration to understand the subnuclear structure of nature.

When an LEP positron and electron collide, they are transformed into a virtual particle, either a Z^0 boson or a photon. The photon or Z^0 almost immediately decay into other particles, which are then detected in the huge particle detectors. During LEP's first physics run of three months commencing September 1989, each detector recorded 30,000 Z^0 particles, establishing the power of the LEP collider for producing Z^0s for the detailed study of the electroweak interaction. In its 11 years of operation, it generated more than 17 million Z particles. Precision analyses confirmed major aspects of the electroweak model, which previously had only been known qualitatively. Measuring the creation and decay of the bosons was a critical test of the electroweak theory.

One of the significant LEP collider findings by the 400 member team, using the Aleph detector, was that the total number of particle families (generations) having light neutrinos was determined to be 2.990 ± 0.015. This number was presented by its spokesperson Jack Steinberger in 1995. The precision of this number was consistent with the electroweak value of 3 (associated with the electron, muon and tau leptons). The number was based on the meticulous measurement of the Z^0 with their decays into hadronic final states as a function of LEP's energy over the range 88 to 95 GeV. The width and height of the Z^0 decay spectra was calculated by the sum of all possible decay possibilities. Why the number of families of the universe was only three remains a mystery today. (In 1968, Steinberger had left Columbia University and joined CERN initially as the Director of Experimental Research in particle physics. In 1988, USA's President Ronald Reagan awarded Steinberger the National Medal of Science.)

The Higgs Boson

Concurrent with measuring the production and decay properties of the intermediate vector bosons, there was an intensive search for a Higgs particle. The Higgs was a part of the electroweak theory needed for self-consistency. No indication of the particle was found when LEP operated in the energy range of the Z particles (1989–1995) and later (1995–2000) when LEP upgraded to 200 GeV. At the close of 2000, LEP was switched off and taken apart to make room in the tunnel for the assembly of the LHC (Large Hadron Collider). The LHC continued the search for the Higgs particle.

Electroweak Problem

The early electroweak theory was based on the concept of symmetries in nature in which the physical properties described remain unchanged under particular transformations such as a rotation or translation. Using this concept, among others, a unified set of equations for both the electromagnetic force and the weak nuclear force evolved into the electroweak theory with an associated electroweak force. To allow the unification of the electromagnetic and weak nuclear forces into a single electroweak force, the Standard Model requires all the carriers of the electroweak force to have the same—or symmetric—zero mass. Yet, unlike the massless photon, which carries the electromagnetic force, the W and Z^0 bosons, which carry the weak force have non-zero masses; the electroweak symmetry of boson masses was thus broken.

Solution

In 1964, two Belgian theoretical physicists, Robert Brout (1928–2011) and François Englert (1932-), in one publication and in another independent publication by Great Britain's theoretical physicist Peter Higgs (1929-) presented a solution. The resolution was a new mechanism that would hide the electroweak symmetry. The Brout-Englert-Higgs mechanism (today referred to as the Higgs mechanism) introduced a new quantum field called the Higgs field. The Higgs boson was its quantum manifestation. The particles that carry the weak force acquired their mass through interactions with an all-pervasive Higgs field. The Higgs mechanism implies that the masses of the particles are dependent on how strongly each particle couples to Higgs bosons—excitations of the Higgs field. These interactions slow down a particle, which is what is meant by inertial mass. It was originally devised to explain the mass of the W and Z^0 bosons but theorists soon found they could extend the Higgs mechanism to explain the mass of all elementary particles. In the Higgs mechanism, particles are born massless (perfect symmetry). The mass giving interaction with the Higgs field hides this initial symmetry.

Although the zero-mass conundrum was solved mathematically years ago, whether the mathematics represented physical reality remained to be tested. The Standard Model, which absorbed the electroweak model, was so successful that the Higgs boson, or something similar was assumed to be present. Finding the Higgs boson, thus became one of the primary motivations for building the LHC so as to study its production and decay properties, proving or disproving, the Standard Model. Although there was little guidance as to the magnitude of the Higgs boson mass, based on searches at LEP it was known to be higher than 114 GeV/c^2. Other considerations limited it to be less than 1 TeV/c^2. Using this knowledge, the LHC was designed to accelerate two opposing proton beams, each of 7 TeV, in the now decommissioned LEP tunnel.

The construction of the LHC was approved in 1995. The total cost of the project was estimated to be 4.4 billion dollars for the accelerator of which 1.1 billion dollars was CERN's contribution to the experiments. To guide the two counter-rotating beams of protons into a collision in the 27 km circular tunnel, the LHC used 1200 superconducting magnets, each weighing more than 25 tons, with each carrying a current of 11 kilo amps, which produced an 8.3 T magnetic field. (One tesla equals 10^4 gauss). The use of the super conducting magnets, cooled by liquid helium, was crucial to keeping the power used to about 20% of that consumed by the Canton of Geneva, when in operation. The collider had four crossing points, around which were positioned seven detectors, each designed for specific types of research. The two major purpose detectors were referred to as ATLAS, a toroidal apparatus, and CMS, a compact muon solenoid.

The LHC, constructed between 1998 and 2008, developed into a collaboration of over 10,000 physicists and engineers and hundreds of laboratories and universities, from more than 100 countries. The ATLAS detector was a partnership among 3000 physicists, from 183 institutions in 38 countries. Being the largest of the detectors in a seven-story chamber, it filled a cylindrical space surrounding its collision region of 25 m in diameter, spreading over a length of 46 m, weighing 7000 tons and containing 3000 km of cable. The CMS detector and partnership was similar to ATLAS. It was a joint effort consisting of over 4000 scientists, representing 206 scientific institutions and 47 countries. The LHC, together with its detectors, represents the largest and most complicated machine ever built by human beings.

By the end of November 2009, the LHC had generated an energy of 1.18 TeV/ beam exceeding the Tevatron's previous record of 0.98 TeV/beam, thus becoming the world's highest energy particle accelerator. On March 2010, the LHC reached 7 TeV by colliding its beam protons together, setting a world record for high-energy collisions. Towards the end of 2010, the ALICE experiment, one of the LHC's first, used the early collisions generated to produce and study matter under the extreme conditions similar to those expected shortly after the Big Bang.

By 2012, data from over 6×10^{15} proton-proton collisions was analyzed. The data was shared worldwide on the LHC Computing Grid that encompassed 170 computing facilities in 36 countries. Data from the ATLAS and CMS detectors was eventually coalesced into a number of potential decay channels expected for the Higgs boson. On July 4, 2012, at a famous CERN seminar transmitted worldwide, spokespersons from each of the independent ATLAS and CMS collaborations presented their results consecutively. Each collaboration found a 5 SD (standard deviation) enhancement at a mass of 125 GeV/c^2 in a graph of events plotted per GeV/c^2 from 100 to 160 GeV/ c^2 mass. The Higgs boson had been found, decaying into pairs of Z^0 bosons and pairs of photons at a mass 133 times that of the proton (0.938 GeV/c^2). From the angular distribution of the Higgs's decay particles into familiar particles, the charge and spin of the original Higgs boson was measured to be charge and spin zero in agreement with the electroweak model.

Over the next few years, the interaction (coupling) strength of the Higgs boson to decay into pairs of muon μ (106 MeV/c^2) and tau τ (1.78 GeV/c^2) leptons, bottom **b** (4.18 GeV/c^2) and top **t** (173 GeV/c^2) quarks, and **W** (80.4 GeV/c^2) and **Z⁰**

(91.2 GeV/c^2) vector bosons was measured and shown to fit the Standard Model to an accuracy of 5–10%. Over this range of mass, from 106 MeV/c^2 to 173 GeV/c^2, the Higgs coupling strength agreed with that predicted for the particle mass. This early data, illustrates that the coupling strength depends logarithmically on particle mass, as predicted. The heavier the particle the stronger is its interaction with the universal Higgs field. No decays to combinations of particles forbidden by the model were found.

In 1993, Leon Lederman and Dick Teresi wrote a popular science book titled *The God Particle,* in part, to promote the search for the Higgs boson (being the God particle). It popularized particle physics and assisted putting the eyes of the world on its discovery in 2012. The popular press called the Higgs boson the God particle because, according to the theory, as explained by Peter Higgs and others, it was the physical manifestation of an indiscernible universal field that gave mass to all matter immediately following the Big Bang. As the universe expanded and cooled, massless particles were formed and interacted with the Higgs field to give them mass, after which they eventually coalesced into the stars, planets, etc.

In 2013, the Nobel Prize in Physics was awarded to Higgs and Englert "…for the theoretical discovery of a mechanism that contributes to our understanding of the origin of mass of subatomic particles and which recently was confirmed through the discovery of the predicted fundamental particle, by the ATLAS and CMS experiments at CERN's Large Hadron Collider."

The Standard Model

The particle discoveries discussed here led to the Standard Model of Particle Physics, which, together with its associated experimental observations permitted physicists to peer deeper into the core of the nucleus and the underlying quantum nature of the Universe than any microscope could.

The Standard Model is based on a quantum field theory that provides expectations, which have been tested successfully in the laboratory. The fields predicted by the model interact through their particles whose existence is established by experiment. This model allows physicists to develop an understanding of the evolution of matter and energy together with the subsequent evolution of the Universe from the Big Bang forward.

In the Standard Model, the basic building blocks of the universe are divided into two types: matter particles called fermions and force carrying particles called bosons. Fermions are the quarks (up, down, etc.) and leptons (electron, electron neutrinos, etc.). Bosons are the force carrying particles (photon, gluon, etc.).

The Standard Model describes three of the four forces (fields) of nature: electromagnetism, the strong force and the weak force. Each force field is carried by an associated particle, called a boson: in the case of electromagnetism, it is the photon; for the strong interaction it is the gluon; and for the weak interaction it is the vector bosons W$^+$, W$^-$ and Z$^\circ$. Interaction with the Higgs boson of the Higgs field gives mass to the particles. The particles transfer force and energy between themselves through the exchange of bosons.

In the Standard Model, the fermions are grouped into three families, each family comprises two quarks and two leptons. The first family includes: the up and down quarks, and electron and electron neutrino leptons; the second family includes, the strange and charm quarks, muon and muon neutrino leptons; the third family includes the bottom and top quarks, tau and tau neutrino leptons. The quarks and the leptons interact with all the forces of nature, including gravity, except the leptons are not affected by the strong force.

Only first family particles occur naturally in nature. The heavier families are created in high-energy collisions, as in cosmic rays or particle accelerators, and they decay rapidly. They are theorized to be present during the early aftermath of the Big Bang, when the universe was extremely dense and hot.

It is not yet understood why there are just three families of particles and, although they are very similar in structure, they have different masses. The Standard Model does not explain these masses. It requires 19 arbitrary parameters and cannot be the final explanation of the elementary particle structure of nature. In addition, it does not include the fourth force of nature, which is gravity.

In an analogy with the Standard Model, physicists have theorized that the gravitational force is also mediated by an unobserved elementary particle called a graviton. Because the gravitational force has a long range, propagates at the speed of light and its source is a field described by Einstein's stress energy tensor, a graviton is likely to be a massless spin two boson. Such a theory would need to reduce to general relativity in the weak quantum field limit and is yet to be developed.

Despite these limitations, the Standard Model resolves many puzzles. It helps explain why only 1/3 the number of expected solar neutrinos were measured first by Davis and later by others, for example. They measured only electron neutrinos, which are the first family kind generated in nuclear beta decay in the sun. In their flight to Earth, they decayed into one of the other two types of neutrinos. In summary, the prediction and measurement of the Higgs boson from the early discovery of the electron represents a major achievement in collective learning, bringing theory and experiment together over the course of the last century.

Chapter 14
Summary

The path to how humans came to know the universe had a birthday 13.7 BYA has been long and convoluted. Knowledge that the universe had an explosive beginning is based on modern day scientific theories and discoveries. These discoveries include the measurements leading to the Hubble Law in 1929, the accidental discovery of the CMB (Cosmic Microwave Background) in 1960, and the development of the SM (Standard Model) by particle physicists that led to the search and discovery of the Higgs boson in 2012.

The Hubble Law articulates the rate at which the universe is expanding; the greater the distance a galaxy is from the Earth the faster it recedes. The CMB is the predicted radiation, left over from the Big Bang, that fits an expected blackbody spectrum precisely. The Higgs boson provides support for the SM's role in explaining the early phases of the Big Bang, which leads to the universe having mass.

This history, begins with the immediate aftermath of the Big Bang and continues through the known development of the universe, Earth and life, and to those who are able to comprehend and write history. The early chapters show how fortuitous it is that we exist, considering the number of mass extinctions. Later chapters focus on the social and economic structures that allowed science to evolve, leading to the understanding of the Big Bang.

Essential to the eventual understanding is the social environment that flowed from the European renaissance, voyages of discovery, scientific revolution and the enlightenment. The seeds for these developments grew from earlier discoveries that began in the river valleys of Mesopotamia and Egypt. They continued through Greece with its subsequent Hellenization of the eastern Mediterranean Sea and westward to Italy, where they were absorbed into the Roman culture. Rome's expansion throughout the Mediterranean region, embraced Greek culture, and was eventually brought north to illiterate Europe.

In astronomy, the Sumerians developed a calendar that was based on 12 months to a year, 24 h to a day, 60 min to an hour, etc. In mathematics, a circle was partitioned into 360°. In high school today, most students take a course in Euclidean Geometry,

Greek is still spoken by the Greeks, and Latin evolved into Italian and other Romance languages. A large part of the English vocabulary begins with Greek or Latin prefixes: auto = self, bi = two, pre = before, etc. All helped form a corner stone of European and later North American civilization and science.

During the Enlightenment, the West began the crucial process of separating the State from the Church. By establishing institutions in universal humanistic ethics, a secular civil society was created. Francis Bacon proposed that knowledge could best be advanced by experiment as opposed to interpreting and refining knowledge from established authorities, like the Bible or Aristotle. The result was an intellectual change produced by clarity of reason and the relegation of faith to a subservient position. This led to enormous growth in understanding nature from studies of the many fields of science that developed. Knowledge gleaned from physics led into chemistry, chemistry into biology, biology into medicine, etc. This evolution had a great impact on increasing the quality, health and wealth of Western life.

The prosperity of the Euro-American social economic engine, emerging from the Enlightenment Period, enabled investments in scientific discoveries leading to understanding the Big Bang and its evolution, which functioned synergistically. The Enlightenment's belief that by understanding the world, the human condition could be improved, proved correct. By the end of the twentieth century, the benefits demonstrated in the West were recognized and copied by much of the world. The tribute-taking elites of earlier centuries, like those of India and China, modified politically to catch up.

An article printed in the April 23, 2001, addition of *Science* journal illustrates this evolution. On February 18, 2021, America's spaceship Perseverance's Rover with its attached Ingenuity helicopter landed on Mars. Shown in *Science* was a selfie picture taken by Ingenuity of its own shadow. It was hovering 3 m above Perseverance on the Martian surface, 288 million kilometers distant from Earth. This achievement represents the first controlled flight of a powered aircraft on another planet. Three months later, China landed its own rover: Zhurong on Mars.

Despite these great achievements made by humans, the story thus far is just the first chapter in humanity's continuing search for meaning.

Epilogue

Parallel to the growth of particle physics, related fields of science have developed and matured, in particular, astrophysics and gravitational physics (cosmology). The fundamental results from particle physics lie at the cornerstone of these disciplines. Several major projects underway in these fields, point to the continued human quest to understand the universe's past and current evolution.

Particle Physics

The number and complexity of the SM's (Standard Model's) components, suggest that a simpler underlying structure of nature must exist. These expectations can be searched for indirectly in the precision studies of the decays of the Higgs particle and directly by exploring energies higher than are available at the LHC (Large Hadron Collider), presently at 13.6 TeV. Several proposals are currently being considered.

Among those, the construction of a new FCC (Future Circular Collider) at CERN has gained significant attention, once the LHC experimental program is completed. The FCC project would continue in the successful path of the LEP/LHC, first by building an electron–positron collider, followed by a high energy proton-proton collider in a new 100 km tunnel. This program would enable measurements in an energy range up to 30 TeV and beyond, which is an order of magnitude higher than possible at the current LHC.

At Fermilab an alternative direction is being pursued. There, a dual project referred to as LBNF (Long-Baseline Neutrino Facility) and DUNE (Deep Underground Neutrino Experiment) is advancing. The project would send an intense beam of muon neutrinos, generated at Fermilab by a new 800 MeV superconducting linear accelerator, to a neutrino detector DUNE located 1300 km distant in the 1480 m deep, refurbished Homestake gold mine in South Dakota. DUNE would comprise 34,000 tons of cooled liquid argon. A smaller detector at the injection site would monitor the

© The Editor(s) (if applicable) and The Author(s), under exclusive license to Springer Nature Singapore Pte Ltd. 2024
T. Sanford, *A Whirlwind History of the Universe and Mankind*,
https://doi.org/10.1007/978-981-97-2674-5

escaping neutrino beam. Data generated would: confirm the SM theory of neutrino oscillations among the three types of neutrinos, be sensitive to new particles and forces, and distinguish any asymmetry between neutrinos and antineutrinos. The latter data could help explain why the Big Bang created more matter than antimatter.

Astrophysics

Complementing these frontiers in particle physics, several studies exploring the formation of the universe and commented on early in this book, dark energy and dark matter, are in process. For example, the recently launched JWST (James Webb Space Telescope) is an international partnership between NASA (National Aeronautics and Space Administration), ESA (European Space Agency) and the CSA (Canadian Space Agency), which is designed to look at how galaxies, including stars and planetary systems, were formed after the Big Bang. With its $10 billion cost, 6.2 tons and 6.2 m diameter primary mirror, the JWST has been equally complex to develop as the LHC.

The JWST is sensitive to a time just 200 million years following the Big Bang. The key to its greater responsiveness is in its ability to detect light in the infrared rather than the optical regime, where the scattering from interstellar gas is less. The JWST allows the expansion rate of the local universe to be measured to a precision of 1% and to be differentiated between its early and later expansion rate. These measurements will add insight into dark energy. By measuring the position and gravitational bending of light (gravitational lensing), the JWST will also survey the distribution of dark matter. Its sensitivity to faint emissions from possible circulating matter around primordial black holes, may assist in discovering the origin of dark matter.

Exploring an earlier time frontier in the development of the expanding universe, is the ground based CMB-S4 (Cosmic Microwave Background-S4) experiment, which is designed to study anisotropies in the CMB emission. CMB-S4 is scheduled to enter into operation beginning 2030. At an estimated cost of 650 million dollars, the CMB-S4 is a joint project funded by the DOE (US Department of Energy) and NFS (National Science Foundation) with Berkeley Laboratory, Fermilab and SLAC, which will all be engaging in essential but different roles.

This massive and extensive experiment is projected to study the CMB created in the Big Bang at 375,000 years, when free electrons were swept up by free protons, which formed hydrogen atoms. Prior to this time, the CMB photons were scattering off electrons and other fundamental particles. It was then that these last representatives of the geometry of the early universe, together with its anisotropies, were imprinted on the CMB's energy, direction and polarization. The CMB measured currently is a result of these last scattering processes.

The CMB-S4 diagnostics will include 21 telescopes located in the Chilean Atacama Desert and at the South Pole, whose dry, cold climates provide an optimum viewing of the Earth's sky. These telescopes will be instrumented with more than 500,000 cryogenically cooled superconducting detectors sensitive to the faint anisotropies imprinted in the CMB. Massive datasets will be required to reduce the statistical uncertainties in the weak data signals that are to be recorded. One of the primary goals is to search for evidence, encoded in the measured CMB asymmetries, of the universe's early expansion (inflation) generated from GW (gravitational waves) that were produced in the Big Bang. That precision will enable discovery for how ripples in the structure of space–time might be connected with quantum mechanics. A secondary goal will be to study dark matter and dark energy on the growth and formation of the CMB anisotropies as measured in the small and large field angular scans at the South Pole and Atacama Desert, respectively. This instrumentation will be used to: catalogue galaxy clusters, gamma-ray bursts via their afterglow, examine general relativity on large scales, and study the Milky Way.

Gravitational Physics

The design of several new ground based GW detectors in Europe are in progress, complementing America's LIGO (Laser Interferometer Gravitational-Wave Observatory) and Italy's Virgo GW detector. They are all based on observing GWs produced when two heavy astro-companions spiral into each other, producing minuscule ripples in the fabric of space–time. The most sophisticated of the new detectors, the Einstein Telescope, is being developed to detect an estimated merger of thousands of black holes per year generated soon after the Big Bang. Theoretically, these combinations may have formed early in the development of the universe, giving rise to dark matter. The detectors operate, similar to the LIGO or Virgo, by sending a laser beam through long tubes and reflecting it off mirrors suspended at the end of each tube. Fluctuations in the beam's travel time, indicate a passing GW has expanded or compressed the tube arms.

The Einstein Telescope is designed with three overlapping arms, Each arm is 10 km long, two and one half times longer than those of the two L shaped arms of LIGO, which substantially increases its relative sensitivity. The arms will have two internal laser systems. One system will be cooled to near absolute zero temperature, providing it with a sensitivity to long-wavelength mergers of very large black holes. Einstein's suggested location, somewhere between the European locations of Maastricht, Liege and Aachen, has a layer of soft soil above bedrock that is ideal for minimizing surface vibrations. The Dutch government has indicated that it will fund half of the estimated construction cost of 1.9 billion euros, if it is built near the Netherland, Germany and Belgium borders. Construction is estimated to start in 2026–2027, with science studies beginning nine years later.

In the mergers of heavy objects, i.e. black holes, the frequency of a developing GW decreases, both with the increasing object's mass and its increasing distance from its final collapse location. On Earth, GWs with frequencies reduced to a few

Hz, which correspond to objects with masses up to a few thousand solar masses, are detectable in the present GW instrumentation. Below that frequency, the Earth's surface lacks stability. To access the milli-Hz and sub-milli-Hz frequencies, instrumentation positioned in space is required. It is this low-frequency region that is the domain of super massive objects, having millions of solar masses located in galactic centers; and where tens of thousands of compact objects in the Milky Way, emit their signals for years and centuries as they rotate around one another, before entering the final few seconds of their collapse.

The space based GW observatory LISA (Laser Interferometer Space Antenna) is being considered to fill this frequency gap. The LISA project is anticipated to be complementary to the existing and future ground based observatories, such as the Einstein Telescope. Physically, LISA is a multiple interferometer comprising three satellites arranged in an equilateral triangle with sides 2.5 million km long, traveling along an Earth-like heliocentric orbit. The distance between the satellites is measured precisely to detect a passing gravitational wave. When a GW crosses the satellites, it alternatively squeezes one virtual arm and stretches the other, causing these distances to oscillate by just a few nm. Unlike terrestrial observatories, which keep their arms locked in a fixed position, LISA keeps track of the satellite's position by counting the number of wavelengths by which their separation changes each second. It will have a sensitivity to measure relative displacements with a resolution of 20 pm (that is less than the diameter of a helium atom) over distances of several million kilometers. Presently, LISA is in its final study phase. It is a major collaboration between the ESA with a consortium of 1500 members and NASA. Costs are expected to approach 2 billion dollars. A launch by the ESA is currently planned for the mid 2030's.

Key Events Discussed

Part I: Beginnings

2. Matter Universe

13.7 BYA   Big Bang

4.6 BYA   Sun formed

4.5 BYA   Earth formed

3. Life

3.5 BYA   Prokaryotes
         Stromatolite fossils

2.5 BYA   Great Oxidation Event
         First Snowball Earth

1.8 BYA   Eukaryotes

750–580 MYA   Last snowball-hothouse cycles

635 MYA   Multi-celled animals
         Ediacaran Period

532 MYA   Sea life expansion—trilobites
         Cambrian Period

416 MYA   Plants expand to land
          Devonian Period

375 MYA   Fish transition to tetrapods—tiktaalik
          Devonian Period

323 MYA   Buried carbon fossilizes into coal
          Carboniferous Period

250 MYA   Pangaea forms and Dinosaurs evolve
          Permian-Triassic mass extinction

66 MYA   Continents grouped as at present but spread farther apart
          Cretaceous-Paleocene mass extinction

23 MYA   Expansion of grasses and apes
          Miocene Epoch

Part II: Humans

4. Human Evolution

4.4 MYA   Hominins
          Ardipithecus lived in trees and walked

2 MYA   Homo erectus
          Hunter-gatherer

300,000–200,000 YA   Homo sapiens
          Humans

70,000–60,000 YA   Modern humans migrated out of Africa to Middle East
          45,000 YA, Reached Europe
          20,000 YA or earlier, Reached North America

5. Agricultural Revolution

14,500 YA (12,500 BCE)   Fertile Crescent:
          Natuffian villages

3500 BCE   Mesopotamia—Sumer
          City-states

2700 BCE   Recorded history—Uruk
        Epic of Gilgamesh

2700–2200 BCE   Egypt's Old Kingdom
        Age of Pyramids

6. Mediterranean Development

2700–1450 BCE   Minoans
        Advanced European civilization

2334 BCE   Akkadian Empire
        Sargon the Great

1600–1500 BCE   Thera eruption
        Santorini formed

1600–1100 BCE   Mycenaeans
        Greek people of Homer

1550–300 BCE   Phoenicians
        Proto alphabet

800–336 BCE   Greeks
        Complete alphabet

509 BCE   Roman Republic founded
        Roman Law

480–404 BCE   Athenian Golden Age
        Foundations of Western Civilization

356–323 BCE   Alexander the Great
        Conquered Persia

323–146 BCE   Hellenistic period
        Greek cultural expansion

58–51 BCE   Julius Caesar subdues Gaul

44 BCE   Julius Caesar assassinated

27 BCE   Roman Empire founded
        Augustus Caesar emperor

166 CE   Antonine Plague
            Smallpox

180   Marcus Aurelius dies
        End of Rome's Golden Years (96-180)

250–270   The Cyprian Plague
            Hemorrhagic viruses

330   Rome moves to Constantinople
        Christianity legalized

476   Western Roman Empire collapses
        Ostrogoth Odoacer becomes King of Italy

529   Justinian Codifies Roman Law

541   Justinian Plague
        Bubonic plague

650   Eastern Roman Empire Loses Middle East
        Renamed Byzantine Empire

7. Europe's Beginnings

481   Clovis—Frankish King
        Merovingian dynasty

800   Charlemagne—First Holy Roman Emperor
        Carolingian dynasty

929   Caliphate of Cordoba
        Ibn Rushd Averroes

962   Otto I— consolidated German Lands
        Crowned Holy Roman Emperor in Rome

1066   Battle of Hastings
         Norman—William the Conqueror

1088   University of Bologna
         Scholasticism

1095   First Crusade
         Returned Jerusalem to Christianity

1215   Magna Carta
       No free man shall be imprisoned except by lawful judgment

1258   Parliament institutionalized
       Royal power constrained

1309   Papacy moves to Avignon

1347   Black Death
       Bubonic plague

1350   Renaissance
       Florence, Petrarch, humanism

1400   Canon
       Gun powder

1450   Gutenberg printing press
       Quickened spread of knowledge

1453   Eastern Roman Empire ends
       Constantinople falls to Ottomans

8. Transition: 1500–1700

1492   Christopher Columbus:
       Discovers America

1517   Martin Luther's theses
       Corruption in Roman Catholic church

1543   Copernican heliocentric model
       Scientific Revolution

1609   Galileo Galilei
       Discovers phases of Venus confirming heliocentric model
       Develops Mechanics

1687   Newton's Laws of Motion
       $F = ma \ ..... \ F = g \, m_a \, m_b / r^2$

1690–1784   Enlightenment
       John Locke (Liberalism), Montesquieu, Voltaire, Denis Diderot

9. Road to Modernity

1764   Textile industrialization
           Hargreaves's spinning jenny

1776   Watt's steam engine
           Breakthrough in power

1776   American Revolution
           Democracy

1789   French Revolution
           Spreads democracy

Part III: Physics

10. Physics: 1700–1900

1787   Chemistry develops
           Antoine Lavoisier discovers elements: hydrogen, carbon, nitrogen, oxygen

1802   Energy and Power defined
           Thomas Young...kinetic energy $mv^2/2$

1808   Atomic theory develops
           Camus Dalton asserts all matter composed of unique atoms

1850   First Law of Thermodynamics
           James Joule...total energy of a system is conserved

1873   Electricity and Magnetism unified
           Maxwell's Equations...light is a electromagnetic wave

11. Physics: Around 1900

1895   X-rays discovered
           Wilhelm Roentgen

1896   Radioactivity discovered
           Henri Becquerel

1897   Electron discovered
           J.J. Thomson

1900   Blackbody radiation explained
       Max Planck proposes light energy is quantized

1905   Einstein's Special Theory of Relativity
       $E = mc^2$

12. Physics: 1900–1950

1911   Nucleus discovered
       Rutherford model of atom

1913   Quantum structure of atom
       Bohr model

1915   Einstein's General Theory of Gravity
       Light bent passing huge masses...black holes predicted

1924   Wave-particle duality
       De Broglie model

1926   Quantum mechanics develops
       Schrodinger Wave Equation

1932   Dirac's predicted anti-electron (positron) found in cosmic rays
       Carl Anderson

1932   Neutron discovered
       James Chadwick

1932   Neutron decays into proton, electron and postulated neutrino
       Enrico Fermi and Wolfgang Pauli

1938   Sun's energy explained by hydrogen fusing into helium
       Hans Bethe

1938   Nuclear fission discovered
       Otto Hahn

1945   Atomic bomb exploded
       Nuclear fission with chain reaction demonstrated

1947   Vacuum is caldron of virtual particles
       Lamb shift helps form basis of quantum electrodynamics

1952   H bomb exploded
       Nuclear fusion demonstrated

13. Particle Physics: 1950–2023

1950   Cosmic pions and muons produced at cyclotrons

1955   Antiproton discovered at the Bevatron
        Emilio Segre and Owen Chamberlain

1956   Neutrinos detected at a nuclear reactor
        Fred Reines and George Cowan

1956   Parity symmetry violated in weak decays
        Chien-Shiung Wu and Leon Lederman

1962   Two different neutrinos discovered
        First suggestion of subatomic particle families

1964   Charge parity symmetry violated in weak decays:
        Helps explain why we live in a matter universe

1968   Only 1/3 calculated fusion solar neutrinos measured
        Raymond Davis

1968   Discovery of hadronic particles in hydrogen bubble chamber at Bevatron
        Luis Alvarez

1969   Quark model explains newly discovered hadrons
        Murray Gell-Mann and Yuval Ne'eman

1974   Charm quark accidentally discovered at BNL and SPEAR
        Gives experimental credibility to quarks

1975   Tau lepton discovered at SPEAR
        Martin Perl

1977   Bottom quark discovered at Fermilab
        Leon Lederman

1979   Gluon discovered at PETRA
        Strong-force exchange particle binds quarks together

1983   Intermediate vector bosons discovered at CERN's SPS collider
        Carlo Rubia and Simon van de Meer

1995   Top quark discovered at Fermilab's Tevatron
        Completes expected quark sequence

1995   Number of neutrinos limited to three
        Jack Steinberger ALEPH collaboration at LEP collider

2012   Higgs boson discovered at CERN's LHC
        Explains origin of mass

Bibliography

Books

Alvarez, Walter, *A Most Improbable Journey-a big history of our planet and ourselves,* W.W. Norton & Company, 2017

Bartlett, Kenneth R., *The Development of European Civilization*, The Great Courses, The Teaching Company, 2011

Bartlett, Kenneth R., *A Short History of the Italian Renaissance*, University of Toronto Press, 2013

Bauer, Susan Wise, *The Story of Science*, W. W. Norton & Company, 2015

Chaisson, Eric, *Epic of Evolution: seven ages of the universe,* Columbia University Press, 2005

Chaisson, Eric J., *Cosmic Evolution: the rise of complexity in nature*, Harvard University Press, 2001

Christian, David, *Maps of Time: an introduction to big history,* University of California Press, 2005

Christian, David, Cynthia Stokes Brown, Craig Benjamin, *Big History: between nothing and every thing, McGraw-Hill Education,* 2014

Christian, David, *Origin Story: A Big History of Everything*, Little, Brown and Company, 2018

Close, Frank, *Theories of Everything*, Profile Books Ltd, 2017

Close, Frank, *Elusive: How Peter Higgs Solved the Mystery of Mass*, Basic Books, 2022

Crosby, Alfred W., *The Measure of Reality: Quantification and Western Society, 1250–1600,* Cambridge University Press,1997

Daileader, Philip, *The Early Middle Ages*, The Great Courses, The Teaching Company, 2004

Daileader, Philip, *The High Middle Ages*, The Great Courses, The Teaching Company, 2001

Daileader, Philip, *The Late Middle Ages,* The Great Courses, The Teaching Company, 2007

Dartnell, Lewis, *Origins: How Earth's History Shaped Human History*, Basic Books, 2019

Davis, James C., *The Human Story: our history, from the stone age to today*, Harper Perennial, 2005

Diamond, Jared, *Guns, Germs and Steel: The Fates of Human Societies*, W. W. Norton and Company, 1999

Dorsey, Armstrong, *The Black Death: The World's Most Devastating Plague*, The Great Courses, The Teaching Company, 2016

Eisberg, Robert Martin, *Fundamentals of Modern Physics*, John Wiley & Sons, 1961

Goldberg, Dave, *The Standard Model in a Nutshell*, Princeton University Press, 2017

Goldman, Steven L., *Great Scientific Ideas That Changed the World*, The Teaching Company, 2007

Halliday, David and Robert Resnick, *Physics for Students of Science and Engineering*, John Wiley & Sons Inc., 1960

© The Editor(s) (if applicable) and The Author(s), under exclusive license to Springer Nature Singapore Pte Ltd. 2024
T. Sanford, *A Whirlwind History of the Universe and Mankind*,
https://doi.org/10.1007/978-981-97-2674-5

Harari, Yuval Noah, *Sapiens: A Brief History of Humankind*, Harvill Secker, 2014

Harper, Kyle, *The Fate of Rome: Climate, Disease, and the End of an Empire*, Princeton University Press, 2017

Hazen, Robert M., *The Story of Earth*, Penguin Books, 2012

Heather, Peter, *The Fall of the Roman Empire: A New History of Rome and the Barbarians*, Oxford University Press, 2006

Knoll, Andrew H, *Life on a Young Planet-the first three billion years of evolution on earth*, Princeton University Press, 2005

Jackson, Davis Jackson, *Classical Electrodynamics*, John Wiley & Sons, Inc, 1963

Lane, Nick, *Oxygen: The molecule that made the world*, Oxford University Press, 2009

Lane, Nick, *The Vital Question: energy, evolution, and the origins of complex life*, W.W. Norton & Company, 2015

Lederman, Leon with Dick Terese, *The God Particle: What Is the Question?,* Houghton Mifflin Company, 1993

Lederman, Leon M., Christopher T. Hill : *Symmetry and the Beautiful Universe,* Prometheus Books, 2004

McEvedy, Colin, *The New Penguin Atlas of Ancient History*, Penguin Books, 2002

McEvedy, Colin, *The New Penguin Atlas of Medieval History,* Penguin Books, 2002

McEvedy, Colin, *The Penguin Atlas of Modern History (to 1815)*, Penguin Books, 2002

Morin, David, *Special Relativity for the Enthusiastic Beginner*, Harvard University, 2017

Perlov, Delia, Alex Vilenkin, *Cosmology for the Curious*, Springer, 2017

Pinker, Steven, *Enlightenment Now: the case for reason, science, humanism, and progress,* Viking, 2018

Plumb, J. P., *The Italian Renaissance*, A Mariner Book, 2001

Pomeroy, Sarah B., Stanley M. Burstein, Walter Donlan, Jennifer Tolbert Roberts, *A Brief History of Ancient Greece Politics, Society, and Culture*, Oxford University Press, 2004

Pross, Addy, *What is Life?: How Chemistry Becomes Life*, Oxford University Press, 2014

Reich, David, Who We Are and How We Got Here: Ancient DNA and the New Science of the Human Past, Pantheon Books, New York, 2018

Scheidel, Walter, *Escape from Rome: The Failure of Empire and the Road to Prosperity*, Princeton University Press, 2019

Shubin, Neil, *Your Inner Fish*, Vintage Books, 2009

Shubin, Neil, *The Universe Within: the deep history of the human body,* Vintage Books, 2013

Spier, Fred, *Big History and the Future of Humanity,* Wiley-Blackwell, 2011

Stanley, Steven M., John A. Luczaj, *Earth System History*, W.H. Freeman Company, 2015

Steinberger, Jack, *Learning About Particles: 50 Privileges Years*, Springer, 2005

Tinniswood, Adrian, *The Royal Society and the Invention of Modern Science,* Basic Books, 2019

Ward, Peter, Joe Kirschvink, *A New History of Life,* Bloomsbury Press, 2015

Weinberg, Steven, *The First Three Minutes*, Basic Books, 1993

Wolfson, Richard, Jay M. Pasachoff, *Physics*, Little Brown and Company, 1987

Articles

Sanford, Thomas W. L., *Trends in Experimental High-Energy Physics,* Technological Forecasting and Social Change 23, 25–40, 1983

Turler, Marc, Complied by, *Hubble misses 90% of distant galaxies*, CERN Courier, p. 15, December 2016

Grant, Andrew,*Two kinds of waves from a neutron-star smashup*, Physics Today, p. 19, December 2017

Kokkinidis, Tasos, *Ancient DNA Analysis Reveals Minoan and Mycenaean Origins*, Nature, Aug 2, 2017

Gibbons, Ann, *Eruption made 536 "the worst year to be alive,"* Science, p. 733, vol 362, Issue 6416, Nov 16, 2018

Lopatka, Alex, *An ancient merger helped form our galaxy,* Physics Today, p. 19, January 2019

Curry, Andrew, *Who Were the First Europeans?*, p. 94–114, National Geographic, August 2019

Narasimhan, Vagheesh M, et al, *The formation of human populations in South and Central Asia,* p. 999 Science, 6 September 2019

Andres-Toledo, Miguel Angel, *Persian Empire Reborn,* National Geographic History, 2020

Mann, Charles C., *How Microbes Write History,* The Atlantic, p. 14–17, June 2020

Robinson, Andrew, *The Race to decipher Egyptian hieroglyphs,* Science, p. 1574, vol 369 Issue 6511, Sept 25, 2020

Manjit Dosanjh, *Targeting tumors with electrons,* CERN Courier, January/February, p. 21, 2021

Lazaro, Enrico de, *Collapse of Late Bronze Age Civilizations Linked to Climate Change,* Sci-News, 2021

Crivellin, Andreas, John Ellis, *Exotic Flavors at the FCC,* CERN Courier, January/February, p 35, 2022

Voosen, Paul, *Impact crater under Greenland's ice is surprisingly ancient,* Science, p. 1076, vol 375 Issue 6585, March 11, 2022

Borrill, Julian, *Exploring the CMB like never before* p. 34, CERN COURIER, March/April, 2022

Cho, Adrian, *U.S. pares back neutrino experiment to beat rival,* Science, p. 10, vol 376 Issue 6588, April 1, 2022

Cartlidge, Edwin, *Dutch dangle bid for gravitational wave detector,* Science, p. 338, vol 376 Issue 6591, April 22, 2022

Pieri, Marco, Guillaume Unal, *The Higgs Boson Under the Microscope,* CERN Courier, p. 40 July/August, 2022

Lazaridis, Iosif, et al., *The genetic history of the Southern Arc: A bridge between West Asia and Europe,* Science, p. 939, vol 377 Issue 6609, August 26, 2022

Bernhard-Novotny, Kristiane, *Counting down to LISA,* CERN Courier, p. 52 September/October, 2022

Naraoka, Hiroshi, et al., *Soluble organic molecules in samples of the carbonaceous asteroid (162173) Ryugu,* Science, p.789, vol 397 Issue 6634, February 24, 2023

Voosen, Paul, *Anthropocene's emblem may be Canadian pond,* Science, p. 114, vol 381 Issue 6654, 14 July, 2023

Philippsen, Bente, *Dating the arrival of humans in the Americas,* Science, p. 36, vol 382 Issue 6666, 6 October, 2023

Printed in the United States
by Baker & Taylor Publisher Services